Wissenschaftliche Beiträge
zur Medizinelektronik

Band 9

Wissenschaftliche Beiträge zur Medizinelektronik

Band 9

Herausgegeben von
Prof. Dr. Wolfgang Krautschneider

Rajeev Ranjan

Memristive Neuromorphic System Design based on ASICs

Logos Verlag Berlin

λογος

Wissenschaftliche Beiträge zur Medizinelektronik

Herausgegeben von
Prof. Dr. Wolfgang Krautschneider

Technische Universität Hamburg
Institut für Integrierte Schaltungen
Eißendorfer Str. 38
D-21073 Hamburg

Bibliografische Information der Deutschen Nationalbibliothek

Die Deutsche Nationalbibliothek verzeichnet diese Publikation in der
Deutschen Nationalbibliografie; detaillierte bibliografische Daten sind
im Internet über http://dnb.d-nb.de abrufbar.

ISBN 978-3-8325-4850-6
ISSN 2190-3905

Logos Verlag Berlin GmbH
Comeniushof, Gubener Str. 47,
10243 Berlin
Tel.: +49 (0)30 / 42 85 10 90
Fax: +49 (0)30 / 42 85 10 92
http://www.logos-verlag.de

Memristive Neuromorphic System Design based on ASICs

Vom Promotionsausschuss der
Technischen Universität Hamburg

zur Erlangung des akademischen Grades

Doktor-Ingenieur (Dr.-Ing.)

vorgelegte Dissertation

von

Rajeev Ranjan

aus

Vaishali, Indien

2019

Betreuer: Prof. Dr. Wolfgang Krautschneider

1. Gutachter: Prof. Dr. -Ing. Wolfgang Krautschneider

2. Gutachter: Prof. Dr. -Ing. Görschwin Fey

Tag der mündlichen Prüfung: 24 January 2019

Abstract

Development of memristor-based device in 2008 has encouraged scientists and engineers to perform brain-inspired cognitive tasks on the hardware level. Memristor is a resistor with memory which has some similarity like a synapse in our brain and can be fabricated in a 3-dimensional array with high density. This allows memristors to be used in parallel processing applications like pattern recognition.

Memristor-based neuromorphic systems have neuron and memristor integrated together as the fundamental unit which works like neuron and synapses in our brain. This unit is repeated in large number and interacts with each other to perform the complex functionalities of the brain. Many researchers have done pattern recognition and other cognitive tasks on simulation level and partially on hardware implementation. Development of full neuromorphic system requires the development of memristor and CMOS neuron circuits with spiking circuit for implementing the fundamental properties of synapses that is Long-Term Potentiation (LTP) and Long-Term Depression (LTD).

Since the continuous switching memristor is in its infant stage, in order to progress the research in the neuromorphic circuit, an On-chip externally programmable emulator has been designed which shows memory effect (pinched hysteresis curve), LTP and LTD on the hardware level. This emulator can be used for implementing any memristive behavior on the hardware level. A double barrier memristor device (DBMD) has been developed by Institute of Electrical Engineering and Information Technology, University of Kiel which shows a continuous change in resistance when the consecutive write voltage is applied. This device has to be integrated with CMOS neuron to develop potentiation and depression. An integrate & fire neuron chip with spiking circuit has been designed and fabricated in AMS350 nm process.

DBMD has been integrated with Neuron ASIC and the measurement shows LTP on hardware level which can be used for a larger network to mimic functionalities of the brain. A LTspice based simulation has been performed with DBMD model and neuron model to train 4 images, each of 4X3 pixels. The simulation shows the unsupervised learning of a memristor-based system. This paves the way toward development of neuromorphic computers.

To my wonderful parents for their motivation, loving wife, Supriya for her constant support and in loving memory of my grandfather.

Acknowledgements

I wish to acknowledge individuals who mentored and supported me throughout my doctoral research. Firstly, I would like to thank my supervisor Prof. Dr.-Ing. Wolfgang Krautschneider for providing me constant motivation during the progress of my research and providing me all possible support to finish my research successfully. I would also like to thank Dr.-Ing. habil. Dietmar Schröder for sharing his knowledge and experience which provided me the right direction for my research. I also wish to express my sincere gratitude to my colleagues at the Institute for Integrated Circuits for their extended long-term support. Finally, I would like to express gratitude to my parents Mr. Tribhuwan Choudhary and Mrs. Kamini Devi for letting me pursue my dreams and my wife Supriya for her constant support during the process.

Contents

Glossary

f	Frequency in Hertz	**FSM**	Finite State Machine
ms	millisecond	**GBW**	Gain bandwidth product
nm	nanometer	**HNN**	Hardware Neural Network
r_{on}	MOSFET on-resistance	**HP**	Hewlett-Packard
AC	Alternating Current	**IEEE**	Institute of Electrical and Electronics Engineers
ADC	Analog to Digital Converter		
AI	Artificial Intelligence	**LDO**	Low Drop Out
AMS	Austria Micro Systems	**LSB**	Least Significant Bit
ANN	Artificial Neural Network	**LTD**	Long Term Depression
ASIC	Application-Specific Integrated Circuits	**LTP**	Long Term Potentiation
		MNIST	Modified National Institute of Standards and Technology
CMOS	Complementary Metal-Oxide Semiconductor		
		MOSFET	Metal Oxide Silicon Field Effect Transistor
CPU	Central Processing Unit		
DBMD	Double Barrier Memristor Device	**MSB**	Most Significant Bit
		MW	Mega Watt
DC	Direct Current	**NMOS**	N-channel MOSFET
DRAM	Dynamic Random Access Memory	**NN**	Neural Network
		NPU	Neuromorphic Processing Unit
FPGA	Field-programmable Gate Array		
		PMOS	P-channel MOSFET

PDK	Process Design Kit	**UHD**	Ultra High Definition Television
SC	Switch Capacitor		
SMD	Surface Mount Device	**VNA**	Von Neumann Architecture
SPI	Serial Peripheral Interface	**WTA**	Winner Takes All
STDP	Spike Timing Dependent Plasticity		

Chapter 1

Introduction

Von Neumann Architecture (VNA) of modern day computer was given by great visionary scientist Von Neumann in 1945. VNA architecture has separate memory and processing unit as shown in figure 1.1. All programs and data in this architecture are stored in memory and communicate through a bus. Each time a new program has to be processed, the processor fetches the data from memory, process it and store it in the memory. This requires a high bandwidth of bus and separate memory to store each program and data. This architecture has a simple, fixed structure but capable of performing any computation without hardware modification. This paved the way towards the development of modern day computers and helped us in developing a modern day digital world. With the evolution of computers, broad application areas emerged which adopted digital computers for faster and efficient computing. Consequently, high-performance microprocessors and data storage were required. Technology scaling resulted in improving the speed of microprocessor by reducing the size of transistor. However, semiconductor scaling has already reached 7 nanometer (nm) technology and scaling of 3 nm is limited by tunneling. Hence, further scaling is not possible due to physical and practical limits [1]. Therefore, digital processors have limitations in improving it's further speed. Moreover, VNA suffers from communication bottleneck due to bus bandwidth and memory wall resulting from CMOS downscaling [1]. These are the challenges of VNA and limits its usability in applications where huge data and high processing is required like in Artificial Neural Network (ANN).

Challenges in VNA and need for highly efficient computing has encouraged scientists and semiconductor industry to develop a brain-inspired neuromorphic computer which is considered to be highly efficient like a biological brain. Human brain architecture is

1

adaptive in nature and has memory and processing at the same location, unlike VNA. Hence, brain-inspired computing on hardware requires neuromorphic architecture and cognitive device which works similar to synapses in the brain. Neuromorphic architecture can be implemented in CMOS circuit but the bottleneck remains in developing a cognitive device (synapse) which can change its property with time. In 2008, HP lab developed a two terminal device which has adaptive resistance and called memristors. This device has a property like synapses of the brain and attracted scientist and engineers for developing neuromorphic computing with this device. Since its development, multiple researchers have developed memristor with different material and architecture and this device has become a key component in neuromorphic computing [2], [3].

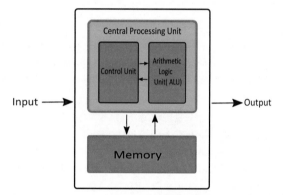

FIGURE 1.1: Von Neumann Architecture

Memristor

The memristor was first postulated by Leon Chua in 1971 [4] where he described memristor as a new two-terminal device which relates two fundamental circuit variables; charge and flux linkage. Chua presents the electromagnetic interpretation of this relationship and shows that the new device has peculiar behavior not found in basic circuit elements like resistor, capacitor, and inductor. In electrical circuits, all the four circuit quantities are linked with each other either by physical law or a physical device as shown in figure 1.2.

Resistor relates voltage & current, inductor relates current & flux and capacitor relates voltage & charge. Chua mentions the missing link in the group which links magnetic flux and introduces a new device which relates these two physical qualities. He termed

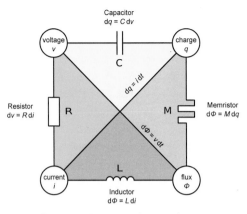

FIGURE 1.2: Two terminal circuit elements

this device name as memristor because it shows the behavior of a resistor with memory. Emulation of this model was done and a non-linear current-voltage relationship was shown in simulation and measurements. Since then, a lot of research has been done to develop such a device which has a non-volatile memory like memristor. A device with flux-charge relation could not be developed till date and it is not expected to be in foreseeable future. However, In 2008, Hewlett-Packard (HP) team under R. Stanley Williams developed a Titanium oxide TiO_2 based two terminal device which has the non-linear characteristic and voltage-current relationship like the device explained by Leon Chua in [4]. This device has similar properties like memristor but was not based on a flux-charge relation. The TiO_2 based device worked on the principle of ion movement when a voltage is applied which eventually changes the resistance of the device [5].

1.1 State of the art and applications

The memristor got attention in scientific community because of it's predicted potential application in non-volatile memory [6], [7], [8], [9] programmable analog circuits [10], [11], [12] neuromorphic circuits [13], [14], [2], [3] and many more. It has opened new possibilities for neuromorphic computers where distributed processing are supposed to be more power efficient and compact.

Non-volatile Memory

3

In last decades silicon based Flash memory has been widely used for non-volatile compact data storage. Flash memory has challenges like high voltage operation and cost effective-production to store ever growing data in present day situation. Research has been going on to develop next-generation flash but it faces technical challenges. Therefore, other type of non-volatile memory like resistive random access memory (ReRAM) [15], [16], [17], phase change memory (PCM) [18], [19], [20], spin-transfer torque random access memory (STT-RAM) [21], [22], [23], magnetic random access memory (MRAM) [24], [25] and Ferroelectric random access memory (FeRAM) [26], [27] have attracted researchers and semiconductor industry as possible alternative for Flash memory. Table 1.1 shows summary of all next-generation non-volatile memory. All possible next-generation non-volatile devices has low read and write voltage compared to convention Flash memory where still more than 10 V is required.

RRAM is getting much attention in industry because it is compatible with conventional semiconductor process [28], [29], [30]. In [30] authors have developed memristor on the silicon substrate and it can be integrated with other CMOS devices like a transistor, resistor, and capacitor on same ASIC. Beside silicon, there is a wide possibility of material which can be used for memristor development. In [31], authors have shown a list of materials which can be used for resistive switching as well as materials which can be used for electrodes. The device can be fabricated in 3-D architecture providing a high density of devices [32], [33]. ReRAM based devices work on the same principle as a memristor. When a positive voltage in applied device change to ON state and when a negative voltage is applied the device change to OFF state as shown in figure 1.3. The devices have different read and write voltage.

Programmable analog circuits

Memristor has attracted analog designer to use this device in programmable analog circuits because it reduces the total area of the circuit compared to the conventional approach. In [10], authors have shown applications of memristor in programmable threshold comparator, programmable gain amplifier, Schmitt trigger and analog filters. The authors have simulated analog circuits with memristor emulator and indicated the successful application in analog design. In [11] author indicates pulse coded precision programmable resistor and simulates the precise change in resistance by changing the pulse width of the clock. Application of memristor has been tested on hardware level in analog applications. In, [12] author uses Titanium oxide-based memristor device to design programmable high pass filter. It indicates the possible use of memristor in analog circuit as it is area efficient, power efficient and less complex compared to the conventional

4

Feature	FeRAM	MRAM	STT-RAM	PCM	ReRAM
Cell size (F^2)	Large, approximately 40 to 20	Large, approximately 25	Small, approximately 6 to 20	small, approximately 8	20 [34]
Storage mechanism	Permanent polarization of a ferroelectric material (PZT or SBT)	Permanent magnetization of a ferromagnetic material in a MTJ	Spin-polarized current applies torque on the magnetic moment	Amorphous/polycrystal phases of chalcogenide alloy	Resistive switching
Read time (ns)	20 to 80	3 to 20	2 to 20	20 to 50	10 [35]
Write/erase time (ns)	50/50	3 to 20	2 to 20	20/30	10 [35]
Endurance	10^{12}	$> 10^{15}$	$> 10^{16}$	10^{12}	10^6 [35]
Nonvolatility	Yes	Yes	Yes	Yes	Yes
Maturity	Limited production	Test chips	Test chips	Test chips	Limited production
Applications	Low density	Low density	High density	High density	High density

TABLE 1.1: Emerging Non-volatile memory summary [28]

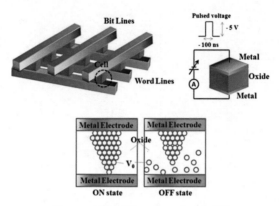

FIGURE 1.3: ReRAM working principle [28]

approach where to program the circuit resistance switching need to be implemented with the digital control circuit.

Neuromorphic circuits

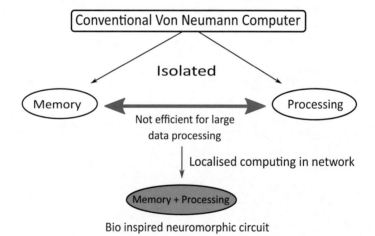

FIGURE 1.4: Conventional Von Neumann computers versus Neuromorphic circuits

Conventional Von Neumann computers have separate memory and processing as shown in figure 1.4. The ever-growing data in present day situation is an issue for even state of the art digital processors to fetch data from memory, process it and store it back in memory. Semiconductor technology has limitation to shrink further because of fundamental and practical limits. Neuromorphic processors may be the next alternative which is bio-inspired and is supposed to be as efficient as a brain. The human brain works at 20 Watts and capable of doing a highly complex cognitive task which is impossible for the modern day computer. Whereas, implementing human-scale intelligence in digital computer will require "more than 10 MW" of power [36] and is therefore impractical [37], [38], [39]. The human brain works much more efficient than digital computers (VNA) because it works on a completely different approach than a human brain. It has a highly dense structure of neurons and synapse which adaptable in nature and remains one of the key components of the biological brain. Figure 1.5 shows biological neuron-synapse and its functional description on hardware. The basic electrically important components of a neuron are cell body, axon, dendrites, and synapse. Every neuron is connected with other neurons through dendrites and each neuron is connected with approximately 10000 other neurons via these dendrites. The interconnect between dendrite and neuron is synapse. It is self-adaptive resistance in an electrical sense which allows neurons to learn. Cell body integrates all incoming currents through these synapses and fires (generates action potential) when it reaches a threshold. Action potential travel along the axon to communicate with other neurons at distance. Implementing bio-logically inspired neuromorphic circuit on hardware requires the development of a cognitive device which is adaptive in nature can be integrated with high density and is adaptive in nature.

Memristor is an alternate device which can be fabricated in 3-D dense structure and it can be part of processing as well as memory. It has opened new possibilities for neuromorphic computers which are supposed to be more power efficient and compact.

This device may give the opportunity to emulate the functionality of the brain, instead of simulating on computers which requires a large amount of area and power. Many research groups are working on brain projects like IBM's Blue brain project, Howard Hughes Medical Institute's Janelia Farm, Harvard's Center for Brain Science etc. Even today not even a rat brain can be simulated because it requires huge power and area.

When the amount of data is huge, which is the present day situation, separate processing, and memory unit approach of von Neumann digital computer is inefficient. A new approach is needed where processing can be done in the distributed manner and memory storage can be localized in a network instead of separate memory.

"Memristor can be made extremely small, and they function like synapse. Using them, we will be able to build analog electronics circuits that could fit in a shoebox and function

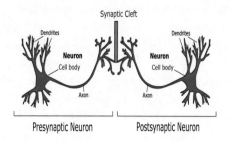

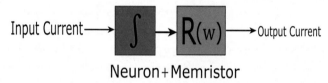

FIGURE 1.5: Synapses and Neuron

according to the same physical principle as a brain. A hybrid circuit containing many connected memristors and transistors could help us research actual brain function and disorders. Such a circuit might even lead to machine that can recognize patterns the way humans can, in those critical ways computers can't - for example, picking a particular face out of a crowd even if it has changed significantly since our last memory of it." [40]

Many researchers have shown successful results of distributed processing and learning behavior with memristors like pattern recognition. In [41] author proposes a design methodology for unsupervised learning of handwritten digits. In this work, pattern recognition has been done on simulation level which shows the pattern recognition even in presence of device variation due to adapting nature of memristor and power of homeostasis. It also shows the self-adjustment capability of the network with different input encoding schemes. Additionally, many researchers have indicated memristor based pattern recognition in simulation [42], [43], [44]. Some authors have developed memristor devices, extracted the measurement data and simulated the device data on hardware for pattern recognition. In [45], Nb_xO_y memristor measurement-based model has been used to train 10 handwritten numbers and system shows recognition rate more than 60 % even with device-to-device variation more than 40 %. Different authors have used different material and its measurement results to simulate the same handwritten data from the MNIST database to achieve pattern recognition. In [46] authors have used $Pd/WO_x/W$

8

device to for pattern recognition and shown the change in memristor value before and after learning. Similarly [14] used TiO_x based device, [47] used $Ag/AgInSb/Ta$-based memristor, [48] used HfO_2 based memristor to simulate memristor device measurement data to perform successful pattern recognition in simulation. Memristor has also been implemented on full hardware with memristor device and FPGA [49]. Neurons, spiking circuits, and homeostasis have been implemented on FPGA and the measurements show successful pattern matching on hardware without any external processing. Memristor-based pattern recognition is less susceptible to noise due to self-adaptation of memristor after training. In [50] author implements $Pr_{0.7}Ca_{0.3}MnO_3$ (PCMO) device based pattern recognition and the system is able to learn 5X6 pixel image with 85 % recognition rate even at 10 % of input noise level.

1.2 Purpose of the work

Challenges of VNA, the requirement of high-efficiency computers, and development of memristor have encouraged the scientific community to develop memristor based neuromorphic processing. Development of neuromorphic processing includes: understanding of memristive device specification for neuromorphic applications, development of memristive devices, it's integration with CMOS circuit, it's evaluation and development of neuromorphic processing model.

Memristor has non-volatile adaptive memory and it has attracted scientific community for developing memristor based neuromorphic circuits. However, there is no commercially available memristor available till now and there is still ambiguity about the memristor behavior best suited for neuromorphic applications. Multiple researchers have developed many different memristive devices with different memristive functions [51], [52], [53], [54], [55] etc. Therefore, a memristor emulator is needed which is compact and externally programmable to implement any memristive function on hardware. Memristor emulator developed by other researchers are bulky and non-programmable ([56], [10], [57], [58]) which limits its usability in neuromorphic circuits. Development of Application Specific Integrated Circuit (ASIC) based emulator will provide freedom to implement a wide variety of memristive function and analyze the impact on learning behavior like the implementation of LTP, LTD and synaptic plasticity. Moreover, two on-chip neurons will provide an opportunity to mimic the combined learning activity of neuron and synapse on hardware.

Brain-inspired neuromorphic computing is considered to be a more efficient approach than digital approach than digital in terms of power and area. Moreover, digital computers have isolated processing and memory which faces technological challenges in dealing high volume of data due to communication bandwidth and the memory wall. Memristor-based neuromorphic processing is capable of handling huge data application like pattern recognition and audio processing due to its distributed computing in the network itself. In memristor based approach, a memristor is part of processing as well as memory. It obviates the need for separate memory and processing unit and emerges as a promising candidate for human-level cognitive tasks applications. Memristor and CMOS neuron based simulation of the neuromorphic circuit shows successful pattern recognition which is a challenge for digital computers when high definition image has to be recognized and processed. Since memristor has distributed memory and processing in the network, architecture can be extended in 3D architecture to recognize and process any high resolution of an image. It shows a wide application area where high-efficiency processors are required and lead towards the development of neuromorphic processors.

Neuromorphic computers are biologically inspired and mimic the functionalities of neuron and synapse in the brain as shown in figure 1.5. Synapse-neuron is repeated in a dense structure in our brain to perform all high-efficiency cognitive task. Therefore, for developing neuromorphic computer, the first step is to develop synapse-neuron unit which can be connected in bigger architecture to perform the human-level cognitive task. Development of NPU requires the development of memristor device, CMOS neuron circuit and memristor-CMOS integration on hardware as demonstrated in figure 1.6.

Development of memristor is beyond the scope of this work and has been developed by our project partner, Institute of Electrical Engineering and Information Technology, University of Kiel. Double barrier memristive devices based on a 4-inch wafer technology have been fabricated which consist of an ultra-thin memristive layer (Nb_xO_y) sandwiched between a Al_2O_3 tunnel barrier and a Schottky-like contact [51]. The presented device indicates a non-filamentary based continuous resistance change. The device shows some basic similarity to synapse in our brain and can be used in neuromorphic applications with CMOS circuits. The neuron ASIC mimics the fundamental properties of a biological neuron; it integrates all incoming current and generates a spike when the threshold of the neuron is reached. Since DBMD requires more than 3 V for changing its conductance, AMS350 nm process has been used for implementing ASIC. Integration of designed ASIC and memristor die develops NPU which is the basic unit of neuromorphic processing capable of learning and storing the memory in a single device. This

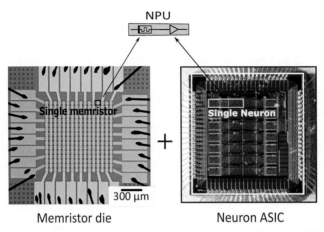

Memristor die   Neuron ASIC

FIGURE 1.6: Memristor die and neuron ASIC integration and development of NPU

NPU paves the way towards the development of neuromorphic computers when arranged in a 3D array.

The author has to thank the German Research Foundation (DFG) for funding the research group FOR2093 (Memristive Components for Neural Systems), Project C2 (Neural Circuits), under which this research activity has been conducted.

1.3   Thesis outline

Chapter two presents the memristor working principle, and its development at nanoscale level. This chapter explains the memristor device developed in HP lab and discusses its measurement results. DBMD has been used in this work to develop memristive learning. It's architecture and advantages in neuromorphic application has been discussed in this chapter.

Chapter three describes the memristor emulator ASIC, its circuit design and measurement results. Since memristor is in its infant stage of development, it is not commercially available till now. Many memristor structure and voltage-current functions have been proposed and each one has different impact on the circuit learning [5], [51], [59], [60] etc. At this stage of research, there is still an ambiguity about the memristive function.

Every memristive function has their pros and cons. In order to understand the specification of memristor for neuromorphic application, an externally programmable memristor emulator hardware is needed. An on-chip memristor emulator has been designed which is compact and can be used in emulating any type of memristive function and can be used for bigger networks as well. The memristor emulator chip has been designed in AMS 350nm process and processing of the memristive function has been done off-chip on a FPGA. This selection allows the implementation of any memristive function on hardware. The emulator has shown a pinched hysteresis curve which is a fundamental behavior of the memristor. It shows two different approaches to implement the emulator on-chip, resistive based and switch capacitive based. It discusses the algorithm to implement the emulator on-chip together with the off-chip processing algorithm and its measurement results.

Chapter four presents the simulation of neuromorphic pattern recognition using the DBMD memristor device model and a CMOS neuron circuit. Pattern recognition simulation of four different patterns in LTSpice simulation software has been shown. It explains the neuron learning algorithm and necessary elements as competition, homeostasis, and variability. This chapter explain the limitations of VNA in implemnting ANN in digital computers and discusses the usability of neuromorphic circuits in high data application like pattern recognition.

Chapter five explains the design of a neuron ASIC on CMOS AMS350 nm process and its measurement results. CMOS neuron circuit had to be designed in such a way that it works like a neuron in the human brain and can be integrated with DBMD to perform pattern recognition which has been done only at simulation level in [45], [14]. This chapter also explains the details of neuron architecture and design considerations of analog building blocks like operational amplifier, comparator, integrator and spike generator circuit used in this neuron ASIC implementation.

Chapter six explains the integration of the CMOS neuron circuit with a real memristor (DBMS). Single memristor device has to be integrated with a CMOS designed neuron chip to develop NPU which can be repeated in 3D array for high effiecient processing like biological neural network. It show the fundamental behavior of learning like Long-Term Potentiation (LTP) which is one of the key components of learning.. The designed neuron chip has an integrate-and-fire circuit (I & F) with an on-chip spike generator circuit. The neuron circuit integrates the current that flows through the memristor at low voltages (read voltage) and generates the potentiation or depression pulses (firing) across the memristor. These pulses result in long-term potentiation or depression of the memristor [61], [62]. The DBMD memristor shows 100 times change in conductance (100 times stronger synaptic contact) with consecutive pulses resulting in learning of memristor (increased firing rate of the neuron). This explains the integration approach and hardware measurement results with memristor retention issue. This chapter shows the learning behavior on hardware level.

Chapter seven presents conclusions and outlook of this work. It proposes neuromorphic processor model based on NPU.

Chapter 2

Memristor

Memristor is a two terminal device which changes its resistance when a voltage is applied across its terminals. The change in resistance depends upon the voltage polarity, voltage level and time duration for which voltage is applied. Memristor resistance is plastic in nature which retains its resistance when the voltage is turned off until next voltage is applied. Memristor has a positive and negative terminal as shown in figure 2.1.

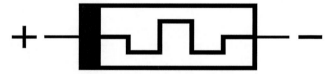

FIGURE 2.1: Memristor Symbol

When the voltage at the positive terminal is higher than the voltage at its negative terminal, the memristor becomes more conducting, while it becomes less conducting when the voltage at its negative terminal is higher. The amount of change in conductance depends on the duration for which a positive voltage is applied. When the voltage is turned off, the memristor freezes its resistance state. The change in resistance is due to the change in position of ions which migrate from one location to other in the device when exposed to a high electric field. The applied electric field has to be high to move ions and to generate a high electric field. To generate a high electric field with low voltage, the interface layer must be very thin in the nanometer scale. So the memristive effect can only be observed in the nanoscale devices.

2.1 How memristor works

The memristor has been built with different materials and different architectures which shows different behaviors. Some examples are the niobium-based (Nb_xO_y) memristor [51], and the Titanium based in (TiO_2) [5] etc. There are two basic principles for memristive devices, oxide drift based [51], [52], [53] and conductive filament based [54], [55].

In the case of the oxide-based memristor when a voltage is applied, oxygen ions change their position and makes the overall conductance change. Titanium dioxide (TiO_2) based device is a case of an oxide-based memristor.

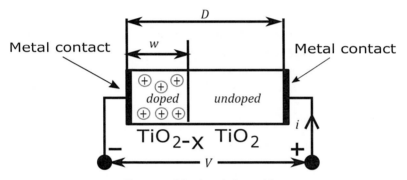

FIGURE 2.2: Memristor device model

The Titanium dioxide (TiO_2) memristor device is a nanometer cube of titanium dioxide (TiO_2) in two layers between two electrodes (metal contact) as shown in figure 2.2. The right layer (TiO_2) has 1:2 titanium to oxygen ratio which acts as an insulator, the left layer has missing oxygen ions which makes this layer oxygen deficient and therefore conductive. When a positive voltage is applied, oxygen ions move from left to right which increases the width of the doped region. Increase in the doped region width (w) eventually rises the overall conductance of memristor. When a negative voltage is applied, some oxygen ions move towards the left side reducing the width of the doped region which then reduces the overall conductance of the memristor as shown in figure 2.3.

This device also shows the special memristive effect that when the voltage is turned off, the oxygen ions do not migrate. The oxygen ions stay where they were and freeze the boundary between the two titanium oxide layers. This phenomenon makes memristor

15

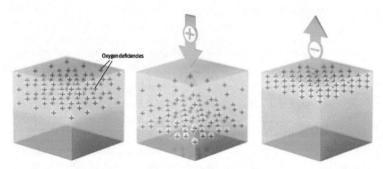

FIGURE 2.3: Memristor ion movement

remember its current state until the next voltage is applied. The device has technological challenges in precisely controlling and fixing the movement of ion and has retention issues [63]. This limited retention of the device may have issues with long-term memory but can be useful in mimicking short-term memory functionalities of the brain.

2.2 Memristor model basic

The memristor has a state variable that represents its resistance. This state variable can be modeled by the equations (2.1) and (2.2).

$$V = R(w) \cdot i \tag{2.1}$$

$$\frac{dw}{dt} = velocity\ of\ w = k_{w,q} \cdot i \tag{2.2}$$

$$k_{w,q} = \frac{dw}{dq} \tag{2.3}$$

where w is a state variable whose derivative denotes the velocity of ion movement which controls the current flow (i), k is the change in width per unit movement of charge and R is the state variable dependent resistance. The state variable *w* depends on the integration of the current flown in the past which is the total charge in this case.

When v(t) is applied across the memristor, the boundary between doped and undoped regions changes. The applied voltage can change x (w/D) between 0 and 1 (figure 2.2). When a positive voltage is applied for a longer time, the doped region increases and resistance reduces leading towards the minimum resistance of R_{on}. When a reverse voltage is applied the undoped region increases which eventually makes memristor achieve its highest resistance state R_{off}. The memristor can be modeled in terms of R_{on} and R_{off} as shown in equation 2.4 and 2.5 and explained in [5].

$$V(t) = (R_{ON} \cdot x + R_{OFF} \cdot (1 - x)) \cdot i(t) \tag{2.4}$$

$$\frac{d(w)}{d(t)} = \mu_v \cdot \frac{R_{ON}}{D} \cdot i(t) \tag{2.5}$$

where μ_v is the mobility of material and **D** is the total device length and **w** is the length of doped region (figure 2.2). Stanley Williams and his team have shown measurement result and pinched hysteresis curve in [5] and brief of the result has been shown in figure 2.4.

2.3 Double barrier memristor devices

The Double Barrier Memristor Device (DBMD) for this project has been developed by our project partner in the University of Kiel for neuromorphic applications ([51], [64]). The device has capacitor-like structure (metal-insulator-metal) as shown in figure 2.5. DBMDs are interface-based device where uniform interface causes uniform change in conductance on contrary to the filamentary based devices ([65], [9], [66], [67]) where memristor shows binary switching. These devices are immune to randomness generated in filamentary based devices ([68], [66], [67]). Some devices has been developed which uses Schottky-like contact for resistance ([69], [70]) change and in some devices electron tunneling property is varied when a memristor layer is in contact with a tunnel barrier ([68], [71], [72]). DBMD combines these two effects into a single device by sandwiching an ultra-thin layer between a tunnel barrier and a Schottky-like contact ([51]) as shown in figure 2.5. This device has many benefits which fits for neuromorphic application. Tunnel barrier of DBMD limits lower resistance state and maximum current of device. This device has better retention than single-barrier device because both the tunnel and the Schottky barrier define chemical barrier to ion migration. The measured retention of

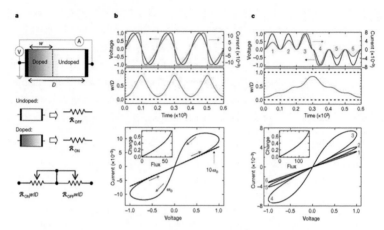

FIGURE 2.4: "**a**, Diagram with a simplified equivalent circuit. V, voltmeter; A, ammeter. **b,c** The applied voltage (blue) and resulting current (green) as a function of time t for a typical memristor. In **b** the applied voltage is $v_0 sin(\omega_0 t)$ and the resistance ratio is R_{off}/R_{on} = 160, and in **c** the applied voltage is $v_0 sin^2(\omega_0 t)$ and R_{off}/R_{on} = 380, where v_0 is the magnitude of the applied voltage and ω_0 is the frequency. The numbers 1-6 label successive waves in the applied voltage and the corresponding loops in the i - v curves. In each plot the axes are dimensionless, with voltage, current, time, flux and charge expressed in units of v_0 = 1 V, $i_0 = \frac{v_0}{R_{ON}} = 10mA$, $t_0 = 2\pi/\omega_0 = D^2/\mu_v v_0 = 10ms$, $v_0 t_0$ and $i_0 t_0$, respectively. Here i_0 denotes the maximum possible current through the device, and t_0 is the shortest time required for linear drift of dopants across the full device length in a uniform field $\frac{v_0}{D}$, for example with D = 10 nm and $\mu_v = 10^{-10}$ $cm^2 s^{-1} V^{-1}$. We note that, for the parameters chosen, the applied bias never forces either of the two resistive regions to collapse; for example, $\frac{w}{D}$ does not approach zero or one (shown with dashed lines in the middle plots in b and c). Also, the dashed i - v plot in b demonstrates the hysteresis collapse observed with a tenfold increase in sweep frequency. The insets in the i - v plots in b and c show that for these examples the charge is a single-valued function of the flux, as it must be in a memristor"

[5]

the device shows better retention for double barrier device compared to Schottky barrier as shown in figure 2.6. This device is a threshold based device contrary to HP memristor device [5] which provides different read and write voltage. As shown in figure 2.5 (top figure), DBMD device allows high current approximately after 1 V which allows this device useful in neuromorphic application where below threshold voltage it act like a resistance and above threshold it changes its conductance. The exact requirement of memristor for neuromorphic application is not yet known and need to be explored but at this stage of understanding DBMD has suits in terms of retention, threshold and current limitations. DBMD will be discussed in detail in section 6.2.1.

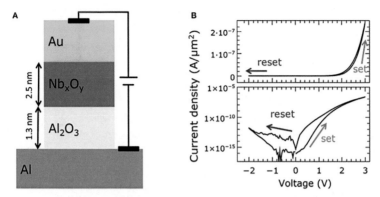

FIGURE 2.5: "Double-barrier memristive device: (A) Schematic cross-section of the Al/Al2O3/NbxOy/Au double-barrier memristive device. (B) Current density J as function of the applied bias voltage. In the upper graph in a linear scale for J was used, while in the lower panel the absolute value and a logarithmic scale was used to better visualize the obtained change in resistance" [45]

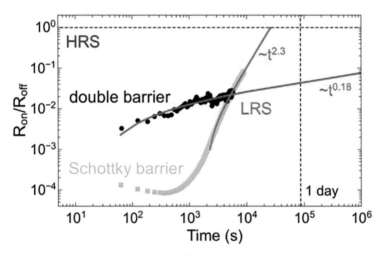

FIGURE 2.6: "Time dependence of the LRS (data points) compared to the HRS (blue dashed line) for the NbxOy/Au Schottky contact of Fig. 2.5 (gray data points) and the double barrier device (black data points). For the readout of the resistance state read pulses of 0.5 V every 60 s were applied. Red lines are data fits used to extrapolate retention times. Inset: retention characteristic for the double barrier device within the first 500 s." [51]

2.4 Conclusion

Analog storage and processing capabilities of memristors have been realized with an oxide drift based memristor device. The operating principle, architecture, and materials used depend on the required application. DBMD has shown better retention than a single barrier device and has a maximum current limit which makes it a better choice for neuromorphic applications. However, to compete with CMOS technology, the memristor has to be power and speed efficient. The operating voltage ranges for these memristor devices extend from -3 V to + 3V which is much higher than state-of-the-art CMOS operating voltage $-/+$ 1 V. This would require scaling down the geometry into the sub-nanometer range which may introduce additional parasitics effects affecting the speed eventually. This will be a great challenge to reduce operating voltage and control the additional parasitic effects simultaneously.

Chapter 3

Memristor emulator and neuron

Memristor emulator is the circuit which mimics the dynamics of a real memristor. Memristive devices have a different domain of application and thus need to have different properties. Many researchers have implemented memristor based circuits in wide applications. However, current day technology has limitations in producing memristive devices due to issues like retention, reproducibility and high voltage operation. Memristors are not commercially available and also, it is not well understood which memristor model is the best fit for neuromorphic applications. This is a present-day challenge and in order to solve this, a compact and reprogrammable memristor emulator is needed in order to implement any memristive function. Many researchers have shown different approaches to design memristor emulator [10], [58], [57]. Table 3.1 shows a comparison of the memristor emulator with the state of the art emulator and real memristor. As seen in the table the implemented emulator presents a bigger memristor range which is limited by the area on the chip, has higher resolution which is limited by the resolution of the ADC and has a higher frequency response which is limited by the overall processing time.

Considering the neuromorphic circuit requirements in mind, an array of memristor emulators has been designed which can be used within a bigger network for the proof of concept of biologically inspired memristive learning. The designed ASIC has a resistive switching and switched capacitor based adjustable resistance implementation. The resistive switch has five memristor emulators with four having conductance ranges from 195 nS to 190 μS (5.2 kΩ to 5.12 $M\Omega$) and one having a conductance range from 4.88 nS to 4.99 μS (200 $k\Omega$ to 204.8 $M\Omega$). The designed emulator has higher memristance range, more discretization steps and has higher operating frequency compared to the

TABLE 3.1: Comparison table of a solid-state memristor and present version of memristor emulator with other emulators

Parameter	Real memristor	**Memristor Emulator**	[10] [Most cited]
Resistance range	Determined by the structure	200 $k\Omega$ <R <204.8 $M\Omega$	50 Ω <R <10 $k\Omega$
Discretization of R	R changes continuously	1024	256
Frequency	Any	<45 kHz	<50 Hz <900 Hz
Response	Determined by the structure	Determined by preprogrammed function	Determined by preprogrammed function
Applied V	Less than the breakdown voltage of structure	0, +3.3 V	0, +5 V or +2.5 V, -2.5 V
Supply V	Not needed	0, +3.3 V	0, +5 V or +2.5 V, -2.5 V
Implementation level	Test die	Designed with ASIC	Designed with commercially available components

state of the art as shown in Table 3.1. Processing has been planned to be off-chip to get the freedom of programmability of any function in ASIC. Besides emulator two neuron circuits have also been designed which integrates all incoming currents and triggers output when a particular threshold is reached. This chapter explains the memristor emulator and neuron circuit. The realization of synapse functionalities used in neuromorphic circuits such as long-term potentiation (LTP), Long Term depression (LTD) and synaptic plasticity has been tested on the emulator.

3.1 Resistive memristor emulator ASIC

A chip containing an array of programmable memristor emulator circuits have been designed and fabricated in AMS 350nm process. Figure 3.1 shows a photograph of the ASIC.

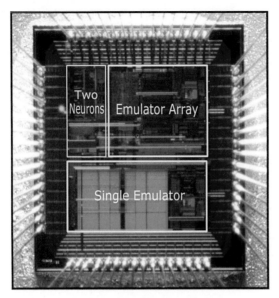

FIGURE 3.1: Emulator ASIC die [61]

3.1.1   Resistive memristor emulator: architecture and design

Memristor emulator has on-chip and off-chip signal processing units as shown in signal flow diagram in figure 3.2 and in block diagram in 3.3. The emulator starts with the start of emulator ASIC and FPGA, ASIC measures the analog voltage across emulator, converts it into serial data, send it for digital processing to FPGA and updates the emulator with a new state variable.

The implemented memristor emulator has an on-chip resistor array with switches that can be controlled digitally, an analog processing block and an off-chip digital signal processing unit able to receive digital data from the chip, process the memristor model and send the setting for the resistor array switches (Figure 3.4) on-chip. External processing can be done either in Matlab or by an FPGA. In this work, an FPGA has been used because FPGA has a faster processing speed than Matlab.

Voltage measurement unit has input buffers which trace the analog voltage across inputs. Scale by 2 block is a differential to a single-ended converter and an attenuator which divides the single-ended voltage by 2. By using a scaling circuit to the voltage

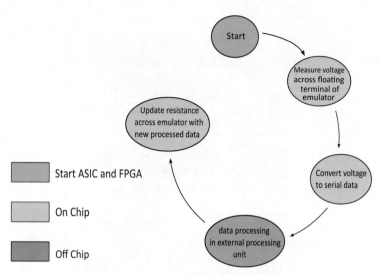

FIGURE 3.2: Memristor Emulator signal flow

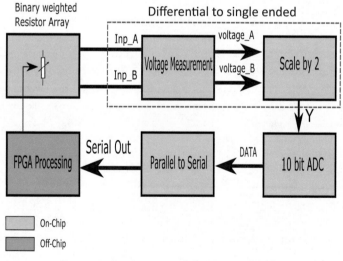

FIGURE 3.3: Resistive switching emulator block diagram

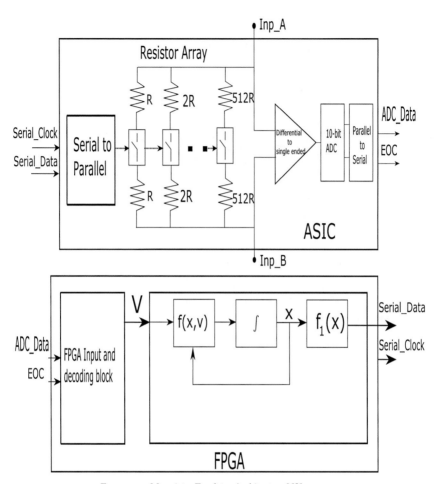

FIGURE 3.4: Memristor Emulator Architecture [62]

measurement, it converts the differential signal at the input to a single ended output **Y** (figure 3.3). In AMS 350nm process the allowed voltage range is 3.3 V. However, each terminal may vary between 0 to 3.3 V which makes the differential voltage to vary between +3.3 V to −3.3 V, this accounts for an overall input range of 6.6 V. In order to accommodate this 6.6 V range in the 3.3 V process, scaling has been done by 2. This single ended signal (**Y**) is converted into a 10 bit parallel digital data by using a 10 bit ADC. The ADC module has been taken from the AMS standard library. Every 10-bit

TABLE 3.2: Building blocks of emulator

Building Block	Block detail	Characteristic
Differential to single ended	Designed	section 3.1.2
Serial to Parallel	Designed	Section 3.1.2
Transmission Gate	Designed	section 3.1.2
Resistor Array	Designed	section 3.1.2
Operational amplifier	AMS 350nm Standard Library	[73]
ADC	AMS 350nm Standard Library	[74]
Parallel to serial	Designed	section 3.1.2

signal is further converted into serial data to communicate with the FPGA. Serial data is received at the FPGA (off-chip) and processed for memristive function implementation. After processing, a digital value is generated corresponding to the required conductance across the emulator input.

3.1.2 Circuit design

On-chip Emulator module has been designed by using the AMS 350 nm standard library IPs and self-designed analog and digital blocks. Table 3.2 shows the list of components used in the memristor emulator.

Differential to single ended output

This circuit has two input buffers to provide a high input impedance, and a voltage scaling circuit at the output as shown in figure 3.5.

This block contains two floating inputs for the emulator (Inp_A and Inp_B) and an output (OUT). The output of the block can be described by the equation 3.1.

$$Vout = 1.65 + \frac{1}{2} * (Inp\_B - Inp\_A) \tag{3.1}$$

where 1.65 V is the virtual ground voltage.

Operational amplifiers OP_AMP1, OP_AMP2, and OP_AMP3 were taken from the AMS standard analog library (A_CELLS). The maximum operating input frequency for this emulator is limited to 45 kHz which is limited by the total processing time in the FPGA and the data conversion time on the chip. Considering this frequency and available operational amplifiers in the library, OP_LN opamp has been found to be best suited because its GBW is 2.3 MHz which is much above the maximum operating frequency. It has much higher input impedance than maximum measurement resistance of emulator which is 200 MΩ. A 100 kΩ and 50 kΩ resistors have been used at the input and feedback respectively to scale the single-ended voltage by half.

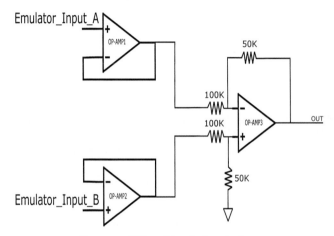

FIGURE 3.5: Differential to single end voltage

The circuit has been simulated for **OUT** signal and it is in accordance with equation 3.1. As shown in Figure 3.6, the terminal Inp_A, in this case, has been kept at contact 1.65 V (red color) and terminal Inp_B has been swept from 0 to 3 V (cyan color). As presented in the Figure 3.6, VOUT (blue color) scales the differential input signal into single-ended signal and scales it by half.

ADC and parallel to serial data conversion

The single-ended output is converted into a 10 bit parallel data by the 10 bit ADC as shown in figure 3.7. This ADC has been taken from the AMS standard library (A_CELLS). Resolution of ADC determines the discretization steps of the emulator. The ADC is operated at 1 MHz and its conversion time is 10 μS. When **power down** signal goes down

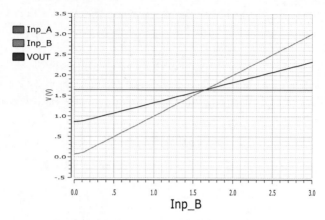

FIGURE 3.6: Differential to single end voltage simulation

and **start ADC** signal goes high, voltage conversion starts. When a conversion is complete, ADC generates an EOC signal which triggers parallel to serial block. This block generates 10 bit serial data which can be sampled in FPGA at the falling edge of a clock as shown in figure 3.8.

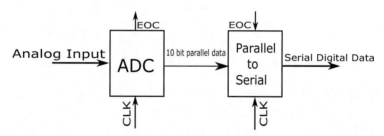

FIGURE 3.7: ADC and serial data output

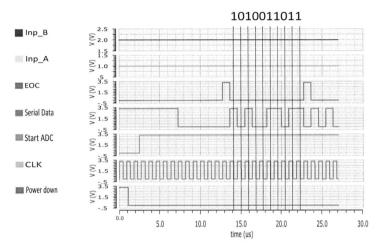

FIGURE 3.8: Differential signal to digital data for 1 V of differential voltage.

The circuit has been simulated for a constant differential voltage of 1 V (Inp_B - Inp_A), the generated digital output is 1010011011 which is equivalent to 667 (0-1023 scale). This way any differential voltage between $+3\ V$ and $-3\ V$ can be encoded by a 10-bit digital data. The digital output is 1111111111 (1023) when the differential input voltage is +3 V and 0000000000(0) when the differential input voltage is -3 V.

Resistor array

The resistor array has been simulated by changing 10-bit digital input from minimum (all bit 0) to maximum (all bit 1). It is expected that the resistance will decrease (increase in conductance) as the digital input changes from lower to higher bits which will lead to a higher current for a constant voltage across the emulator.

Figure 3.9, shows the stepwise current change when the digital input changes the resistor array, this current is given by:

$$I = V * G(Digital\ input) \tag{3.2}$$

where $G(Digital\ input)$ is the digitally controlled conductance of emulator, V is the applied voltage and I is the measured current. Applied voltage in this simulation has been kept constant at 3.3 V and the current passing through the emulator has been measured. The minimum measured current (16.1 nA) and maximum current (16.5 μA) shows the minimum and maximum conductance of 4.88 nS to 4.995 μS for the applied voltage of 3.3 V. This is the minimum and maximum memristance range of emulator. 3.9 shows conductance of 2.2 μS for digital input of 0110111010 and similarly, any other conductance of memristor can be realized on-chip by changing the digital input which is received from FPGA.

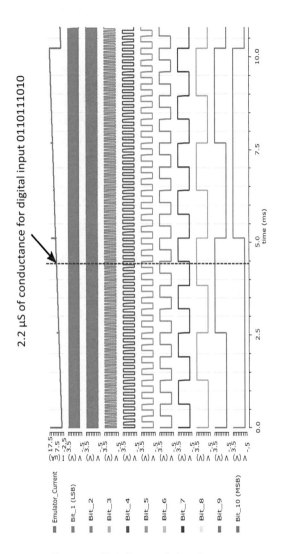

FIGURE 3.9: Digitally controlled conductance

Transmission gate

A transmission gate is used between the resistors to connect or disconnect them. Transmission gate has been designed considering the resistance between which it is connected. The minimum emulator resistance is 200 kΩ so the resistance of the transmission gate has to be at least 100 times less (i.e. 2 kΩ) and the off resistance has to be at least 100 times higher than the maximum resistance in the array. The maximum resistance in the array is 200 MΩ, so the off resistance of the transmission gate is in the range of 475 GΩ which is much bigger than the maximum resistance. ON resistance of MOSFET device depends on the overdrive voltage, W/L and gate oxide capacitance as shown in equation 3.3 for V_{GS} - $V_{TH} >> V_{DS}/2$.

$$RON = \frac{1}{\mu \cdot C'_{ox} \cdot \frac{W}{L} \cdot (V_{gs} - V_{th})} \tag{3.3}$$

A rise in temperature decreases V_{th} which tries to reduce ON resistance but with the rise in temperature mobility decreases and it dominates in increasing ON resistance. With a decrease in C'_{ox} (oxide capacitance normalised to the area), RON also increases and it happens when a device is in a Slow-Slow (SS) corner of the die. So it is mandatory to check the high temperature and SS corner. The transmission gate is simulated for typical and high-temperature SS corner and shows the maximum ON resistance of 2.85 kΩ which is acceptable for this case. Transmission gate has been simulated for OFF state for the voltage across its terminals between 0 to 3.3 V and the transmission gate shows OFF resistance of 475 GΩ which is more than 1000 times higher compared to emulator maximum resistance (200 MΩ)

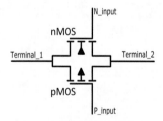

FIGURE 3.10: Transmission Gate

Serial to parallel converter

Serial to parallel converter is used at the input of chip which receives the digital data from FPGA. It has 4 input and 20 output pins as shown in figure 3.11. This block is in the reset state when global reset is low. As soon as the global reset is high this block activates and starts responding the input signals **New Sample**, **Serial_Data_In_Clk** and **Serial_Data_In**. As soon as the new sample signal is received (signal goes low), at each clock edge it samples the data signal and stores it in a 10-bit register. This 10 bit stored data is inverted in order to get the 10 set of complementary data for PMOS and NMOS of the transmission gate (switch). Functional implementation of this block has been done in Verilog and layout implementation has been done with Encounter (Cadence).

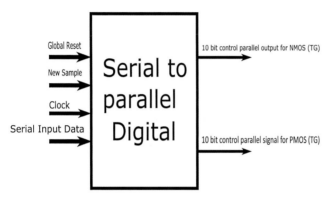

FIGURE 3.11: Serial to parallel converter

3.1.3 Multiplexed four emulator

Emulator ASIC has array of 4 emulators with low resistance range ($10\,k\Omega$ to $10.2\,M\Omega$) compared to single emulator explained in 3.1.1. Each of these emulators can be accessed simultaneously and accessed independently to integrate with neurons. Emulator array has 4 set of floating terminals (Emul_[0:3]_A and Emul_[0:3]_A) for each emulator, 2 selection bit (S0 and S1) as shown in figure 3.12.

Working principle of emulator array remains the same as explained in 3.1.1, however, to reduce area on-chip signal multiplexing has been done. Same analog processing block has been shared by all emulator which samples analog input by selecting selection bits as shown in architecture in figure 3.13. Similarly, input digital signal is multiplexed and used to access individual emulator.

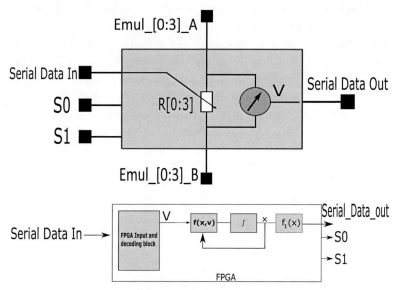

FIGURE 3.12: Multiplexed Four Emulator Architecture [61]

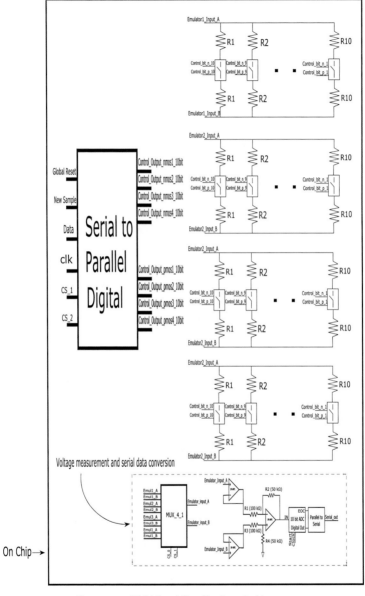

FIGURE 3.13: Multiplexed Four Emulator Architecture

3.1.4 Memristor model implemented on resistive emulator

Various memristor models have been proposed and fabricated till now. Every one of these models has their benefits and limitations [75]. In order to test a memristor model, a state variable dependent conductance model has been implemented in this work which relates current and voltage described by equation 3.4 [61]

$$I(t) = G(x) \cdot v(t) \tag{3.4}$$

where $I(t)$ is current, $v(t)$ is voltage and $G(x)$ is the conductance dependent on the state variable (x). The conductance is given by:

$$G(x) = G_{min} + f_1(x) \cdot (G_{max} - G_{min}) \tag{3.5}$$

where G_{max} and G_{min} are the maximum and minimum conductances fixed on-chip, $f_1(x)$ is a function of the state variable x, which can be programmed externally. Equation 3.4 and 3.5 shows the change of conductance between G_{max} and G_{min} when a voltage is applied across the memristor. The applied voltage leads to a movement of ions in the memristor which eventually produces a permanent change in the resistance. The function $f_1(x)$ in equation 3.5 is a state variable dependent function which can be programmed to be a linear or non-linear function depending on the type of memristor model need to be implemented. In the scope of this work,

$$f_1(x) = x \tag{3.6}$$

was programmed, assuming a linear dependence of conductance on the state variable.

The state variable x which determines the current conductance state changes with the total amount of current flown in the past through the memristor and has been modeled by [61]

$$\frac{dx}{dt} = f(x, v) = k \cdot G(x) \cdot v(t) \cdot f_w(x) \tag{3.7}$$

where $f_w(x)$ is a window function modeling the non-linear dopant drift and limiting the state variable in the range of 0 to 1 [76]. The rate of change of the state variable depends on the present conductance, applied voltage, and the window function. The window function used in this work is proposed in [77] and described by:

$$f_w(x) = 1 - (2x - 1)^2 + \delta. \tag{3.8}$$

36

The window function is used to determine the window of state variable between 0 and 1. The model has a locking problem and when the device is on extreme high conductance or extreme low conductance state, it may get locked there forever even though we apply a reverse voltage. A small parameter δ has been added to (3.8) to avoid the locking problem at boundary [76] which can be considered as noise, the value chosen (0.0003) is negligible compared to the maximum value of $f_w(x) = 1$.

Apart from all variables, every memristor has a technology constant which will depend on the material used and device structure. In (3.7), k is the technology dependent constant which has been taken from the model of [5] can be determined by equation 2.3. $k_{w,q}$ in equation 2.3 is the rate of change of width with the movement of charge. Consequently, k can be defined as the inverse of total mobile charge (total mobile charge = $\frac{1}{k}$) in device by equation 3.9

$$k = \frac{1}{k_{w,q}} \cdot \frac{1}{D} == \frac{\mu}{D^2 \cdot G_{max}}, \tag{3.9}$$

where μ is the dopant mobility, D is the device length and G_{max} is the maximum conductance of the device. All the constants for this model have been taken from [5]. As explained earlier, the conductance of the memristor is a state variable function and it is dependent on the current flown in the past. Considering the model discussed above the model can be first implemented in Matlab using a Simulink model of the designed ASIC and processing performed in Matlab. Taking the parameters, $\mu = 10^{-11} \ m^2 s^{-1} V^{-1}$, $D = 10 \ nm$, $G_{max} = 4.99 \ \mu S$ and $G_{min} = 4.88 \ nS$, the full system has been simulated in Matlab, and current and voltage signals have been plotted. The simulation shows the model characteristic pinched hysteresis curve as shown in figure 3.14. The curve has been drawn for the applied differential sinusoidal input voltage of 6 V_{pp} (peak to peak voltage) and initial conductance ($G_{initial}$) of 3.41 μS for four different frequencies 20 Hz, 35 Hz, 75 Hz and 400 Hz respectively.

The hysteresis curve shown above is dependent on the input frequency. It is understandable from the fact that at low frequencies the memristor finds more time to move the ions which eventually leads to the higher change in the conductance. At very high frequency the memristor behaves like a resistor since it has a reduced time to move the ions, under this circumstance the memristor conductance change is so small that it acts close to a fixed resistor.

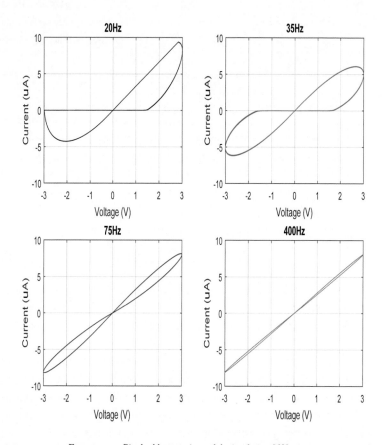

FIGURE 3.14: Pinched hysteresis matlab simulation [62]

3.1.5 Data processing on FPGA

As described before, the memristor emulator ASIC requires an external digital processing unit in order to perform all the computations required for a specific Model. An FPGA was chosen for implementing the processing unit. The model for the memristor is then implemented in Verilog language. The system was designed in order to maximize the parallelism of functions and to provide constant timing. This digital unit periodically requests the digital value of the voltage at the ASIC inputs while at the same time computes the state variable and updates the ASIC conductance. The processing unit consists of two major blocks: The communications module and the Digital Signal Processing module. Data processing on FPGA was designed by the support of Pablo Daniel Mendoza Ponce.

Communication module

The sub-blocks on the communications module were designed to work independently from each other and from the DSP unit. The figure 3.15 shows the main components of the communications module and the interconnections to the signal processing block. As shown, the serial clock is derived from the main FPGA clock and, apart of being sent to the ASIC, is used in the rest of the blocks for synchronization. The data transmission and reception are controlled by two separate state machines that run in parallel.

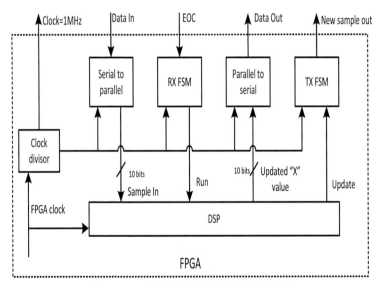

FIGURE 3.15: Top block design in the FPGA [62]

The data transmission FSM is kept locked on its initial state until the DSP block requests an update of the state on the ASIC. The transmission block then loads into the Parallel-to-Serial unit the new 10 bits data present at the output of the DSP. Then the New-Sample-out pin is set to low and the bits are shifted to the Data-out in on each falling edge of the serial clock. The data reception module is triggered by the rising edge of the EOC signal from the ASIC. Once started, the data reception module captures the data bits on the falling edge of the serial clock until it received all the 10 bits representing the voltage at the ASIC input. These 10 bits are then fed to the DSP block together with a signal indicating that a new value is present.

Digital signal processing unit

The data processing was implemented as a custom digital processing unit. The architecture of the processing unit is presented in figure 3.16. The core of the unit is a floating-point unit based on the IEEE 754 standard. This unit is able to compute addition, subtraction, multiplication, division, comparison and conversion between integer and floating point formats. A set of registers is used to store incoming data (voltage sample), constants (e.g. limits for conductance), intermediate results (e.g. computed conductance) and final values (state variable x). Also, in order to reduce the processing time and also reduce the complexity of the module, some memory blocks are dedicated to store precomputed functions (e.g. exponential function of the state variable, e^x) which are indexed by the integer value of the input value (voltage sample) or the memristor state variable x. This generic structure can allow the implementation of different models on the system.

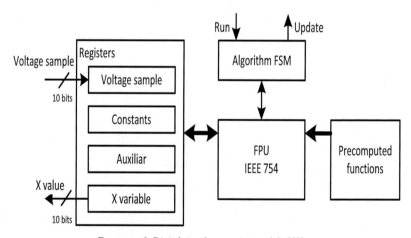

FIGURE 3.16: Digital signal processing module [62].

Finally, the model is coded on the unit as an FSM. Figure 3.17 shows the state diagram for the model used in this research. At start-up or after a reset the FSM initializes the registers of the DSP unit in order to be ready to start the computation of the model. Each computation cycle is triggered by a positive pulse on the Run input (generated by the data reception unit). As a first step, the FSM captures the 10 bits input value and computes the analog voltage value. Then the function $f(x, v)$ (equation 3.7) is computed by using the voltage value previously obtained. Next, the system computes the discrete integral by equation 3.10 and simplified for implementation in equation 3.11.

$$Q_n = Q_0 + Ts \cdot \sum_{k=1}^{n} f(x_{k-1}, v_k) \tag{3.10}$$

$$Q^+ = Q^- + f(x, v) * Ts \tag{3.11}$$

were Ts is the sampling time (time between voltage samples from the ASIC) and, Q^+ and Q^- (initialized to zero) are the updated and present values of the variable Q respectively. Once computed, the new value of Q is compared against the threshold value X_{th}. This threshold value is defined by

$$X_{th} = \frac{1}{k(2^n - 1)} = \frac{D^2 G_{max}}{\mu(2^n - 1)} \tag{3.12}$$

were n is the length of variable X in bits which, in this work, is 10 bits. This threshold represents the value of X that makes a change of one bit on the discrete value of the memristor state variable X. If the magnitude of Q is above the threshold value, then the value of X is updated and the update flag is triggered. Also, in the case of an update of X, the system computes and updates the conductance value (3.5). At the end, the FSM moves to the idle state until a new voltage sample is received.

It is important to indicate that during initialization, the processing unit loads on its registers the constant values for G_{max}, G_{min}, k, Qth and Ts, also a look-up table is loaded in memory containing the pre-computed values for the function $f_w(x)$. The look-up table for $f_w(x)$ is indexed by the integer value of X [62].

3.1.6 Emulator measurement results

The memristor emulator test has been done with the FPGA processing unit to achieve the measurement result same as done in simulation (figure 3.14). The measurement setup for this test is shown in 3.18. As shown in the measurement setup, two sine wave

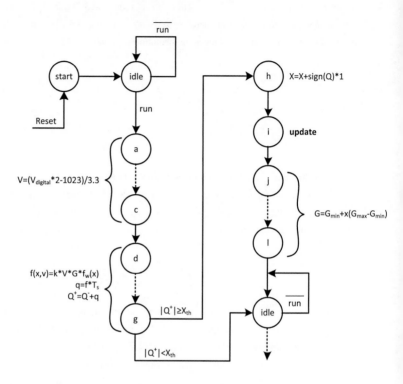

FIGURE 3.17: Data processing unit state machine [62]

signal is applied across the memristor and resistor. The applied differential signal across memristor and resistor is measured at the first channel of oscilloscope and voltage across fixed resistor R is measured across the second channel of the oscilloscope. An isolation amplifier has been used to measure the differential voltage across fixed resistor R and this is independent of universal ground. Later this isolation amplifier has some DC offset. In order to remove this DC offset a DC blocker has been used. This signal measures the voltage across the fixed resistor which eventually gives current across Memristor. Resistance R has a very small value of 100 $k\Omega$ which is very small compared to the maximum resistance of the memristor ($204.8\ M\Omega$) and it impacts the measurement result negligibly.

In oscilloscope, the measured channel 1 and channel 2 can be plotted in X-axis and Y-axis resulting in hysteresis curve shown in fig. 3.19.

The memristor emulator ASIC has been tested with a differential sine wave signal of +3 V to -3 V as shown in figure 3.18 and $G_{initial}$ of 3.41 μS (Programmed on FPGA) for the frequencies of 20 Hz, 35 Hz, 75 Hz and 400 Hz. As seen in the measurement results, and explained in [61] the memristor behaves like a resistor at 400 Hz and it has a high hysteresis at a low frequency of 20 Hz. The measurement results show good agreement with the Matlab Simulink simulation shown in figure 3.14 and 3.19. This shows that this emulator can be used for implementing any memristor model to study the behavior of memristor. The corresponding time-domain waveform of voltage and current are also displayed in figure 3.20 alongside the main window. The range of frequency depends on the model and constants used and can extend up to 45 kHz, which is limited by the sampling period of the ASIC of 11 μs where it takes 10 μs for sending new analog sample and 1 μs for signal processing in FPGA.

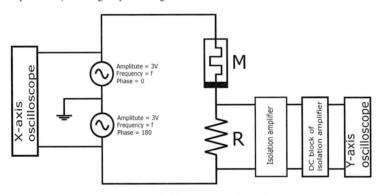

FIGURE 3.18: Measurement Setup [62]

3.1.7 Multiplexed four emulator measurement

FPGA processing of the state variable has been done for each memristor in the same way as explained in subsection 3.1.6. Figure 2 shows the pinched hysteresis curve between current and voltage of each memristor for the applied sinusoidal voltage of +3V to -3V. The memristor has the state dependent resistance which can be measured by current

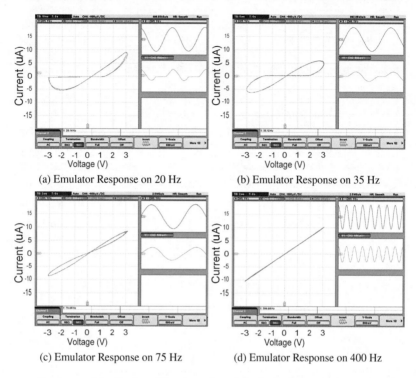

(a) Emulator Response on 20 Hz   (b) Emulator Response on 35 Hz

(c) Emulator Response on 75 Hz   (d) Emulator Response on 400 Hz

FIGURE 3.19: Oscilloscope screenshots of emulator measurement on different frequency. Top right subwindows in each schreenshot shows voltage and current measurement. [62]

and voltage across the emulator. Sine wave voltages of different frequencies have been applied to the four memristors on the chip. Figure 3.20 displays the current/voltage traces of the 4 memristors working simultaneously on the chip, showing the familiar hysteresis curves.

3.2 Switched capacitor based four memristors of resistor range $1\,\mathrm{M\Omega}$ to $7\,\mathrm{G\Omega}$

Memristor is supposed to be low power because of its high resistance of operation. The memristor which has been fabricated in Kiel university has been expected to be in the

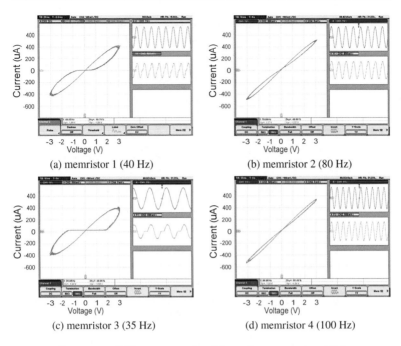

(a) memristor 1 (40 Hz)   (b) memristor 2 (80 Hz)

(c) memristor 3 (35 Hz)   (d) memristor 4 (100 Hz)

FIGURE 3.20: Osilloscope screenshot of 4 memristor measured on ASIC

range of $1\,\text{M}\Omega$ to $7\,\text{G}\Omega$. In order to emulate this range of resistance range in low chip area switched capacitor approach has also been designed.

The architecture of this approach includes non-overlapping clock, switch capacitor circuit and voltage measurement and serial data conversion block as shown in figure 3.21. Voltage measurement and serial data conversion remain the same as explained in chapter 3. The Chip contains 4 such switched capacitor circuit and each one has an independent input clock which determines switched capacitor resistance.

3.2.1 Switched capacitor resistance

Switched-capacitor resistance contains one capacitor (C_S) and two switches S1 and S2 as shown in figure 3.22. Switch S1 is connected with a clock frequency (f) and S2 is connected with an inverted non-overlapped clock of the same frequency (f). Each switching cycle transfers a charge **Q** from input to out at the frequency f.

45

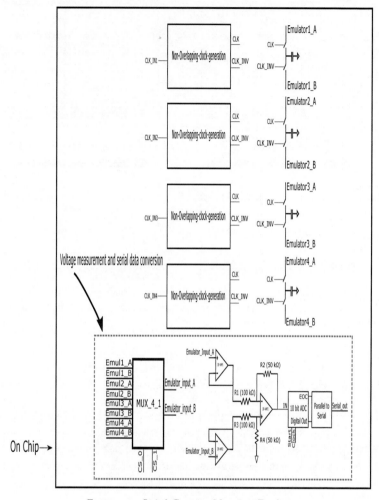

FIGURE 3.21: Switch Capacitor Memristor Emulator

When S1 is closed and S2 is open, a charge stored on capacitor C_S is given by

$$Q_{input} = C_S * V1 \tag{3.13}$$

46

When S1 is open and S2 is closed, fraction of this charge is transferred out of capacitor. The charge remained in capacitor is

$$Q_{output} = C_S * V2 \tag{3.14}$$

The amount of charge transferred from input to output is

$$q_{tx} = C_S * (V1 - V2) \tag{3.15}$$

This charge is transferred at rate of frequency f, So current through circuit is

$$I = q_{tx} * f = C_S * (V1 - V2) * f \tag{3.16}$$

Equivalent Resistance $R_{equivalent}$ is

$$R_{equivalent} = \frac{(V1 - V2)}{I} = \frac{1}{C_S * f} \tag{3.17}$$

As shown in equation 3.17 , $R_{equivalent}$ is inversely proportional to capacitor (C_S) and frequency (f). For a fixed capacitor (C_S), by changing the input clock frequency resistance can be varied. It takes a much smaller area on silicon compared to fixed resistance approach as explained in chapter 3. The non-overlapping clock remains the most critical block of switched capacitor design which will malfunction circuit if the delay between the clocks are less or the clocks are overlapping.

3.2.2   Non-overlapping clock

Non-overlapping clock as shown in figure 3.22 can be generated from single clock as shown in figure 3.23. The amount of delay (δt in figure 3.24) between the non-overlapping clocks are set by delay of nand gates, inverters at the output of nand gate and capacitor (200 fF). When input clock transits from high to low, **delay1_p** signal transits to high delayed by the gate delay of nand gate. **delay2_p** signal is inverted signal of **delay1_p** and capacitor of 1.2 pF introduces delay due to capacitive discharging delay. **delay3_p** signal is the overall delayed signal which is fedback to input of nand gate as shown in figure 3.24. At output a clock buffer is used to provide driving strength of $\phi'1$. Buffered

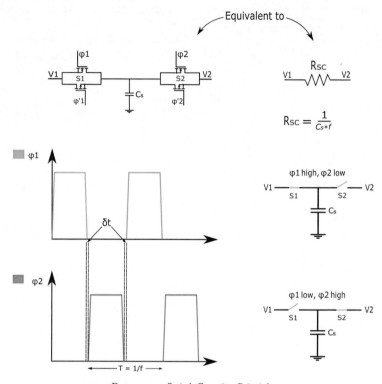

FIGURE 3.22: Switch Capacitor Principle

output clock is the inverted signal of **Input clock** with a delay of 1.83 ns as shown in figure 3.24. When **delay3_p** signal transits from low to high, input **B** of nand (N2) is already at high because **Input clock** is at low voltage level. Since both terminal are at high, nand output transits to low as soon as **delay3_p** goes high. **delay3_p** and $\phi'2$ signal are same signal as **delay1_p** with an additional delay of 1.38 ns which is introduced due to gate delay and capacitor in path.

During transition of input clock from low to high, N2 output switch from low to high because the input B goes low after delay of 1.8 ns the **delay3_p** goes high and this triggers $\phi'1$ to switch from high to low after delay of 1.16 ns as shown in figure 3.24. $\phi1$ is inverted signal $\phi'1$ and $\phi2$ is inverted signal of $\phi'2$. Both signals can be used as input for transmission gate switches in switch capacitor circuit as shown in figure 3.22.

48

The non-overlapping clocks $\phi1$ and $\phi2$ should never coincidence because it will make a short circuit between **V1** and **V2** (figure 3.22). Due to process variation on silicon, a clock may have a different delay and it may generate overlapping clocks instead of non-overlapping clocks. In order to ensure the delay between clock, Monte Carlo analysis has been done for process and mismatch variation. The simulation shows a standard deviation of 157 pS and the minimum delay of 1 nS (@ 25°C and at the typical voltage of 3.3 V) as shown in figure 3.26. The delay will be minimum when the supply voltage is maximum (3.6 V) and the temperature is minimum (-50 °C) because at this state MOSFET devices will be fastest. Monte Carlo analysis has been done at a minimum temperature and maximum voltage and the minimum delay has been found to be 811 pS and standard deviation 110 pS as shown in figure 3.27. In all process, voltage and temperature variations the two clocks will remain non-overlapping.

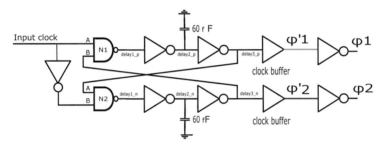

FIGURE 3.23: Non overlapping clock schematic

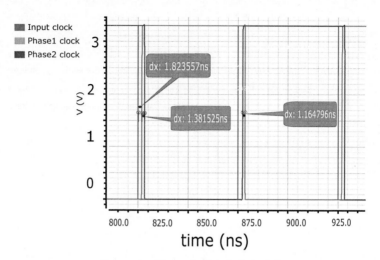

FIGURE 3.24: Non overlapping clock simulation

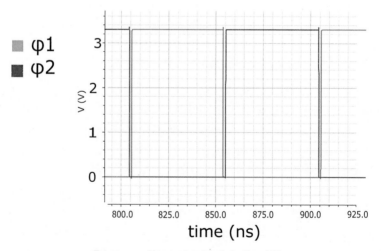

FIGURE 3.25: Non overlapping clock $\phi1$ and $\phi2$

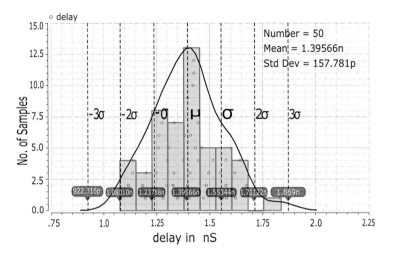

FIGURE 3.26: Non overlapping clock Monte Carlo 27°C and 3.3 V supply

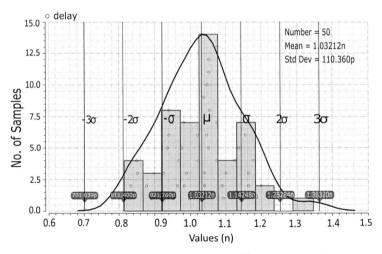

FIGURE 3.27: Non overlapping clock Monte Carlo at -50°C and 3.6 V supply. Device will be fastest at lowest temperature and at maximum voltage inducing minimum delay. Simulation shows minimum delay of 701 picosecond at -50°C and 3.6 V supply which is acceptable for this case.

3.2.3 Switched capacitor simulation and realization of resistance

Switched capacitor circuit shown in figure 3.22 has been simulated for 10 MHz of clock signal. Clock is applied at **input clock** pin of non-overlapping clock circuit (figure 3.23) which generates non-ovelapping clock for transmission gates **S1** ($\phi1$ and $\phi'1$) and **S2** ($\phi2$ and $\phi'2$). When **input clock** is high, $\phi1$ is high and $\phi'1$ low which turns on transmission gate S1. This allows charging of capacitor C_S by a spike current of 118 μA in approximately 1.2 nS as shown in figure 3.28. Current is high when S1 turns on and as soon as the capacitor voltage (Voltage (Cs)) reaches to maximum voltage (3.3 V) the current reduces to approximately zero current. This charging cap stores the charge of 28.71 femtocoulomb as explained in equation 3.13 . This stored charge is transferred to V2 terminal when S1 is OFF and S2 ON ($\phi2$ high and $\phi'2$ low) as shown by the negative current in figure 3.28. The average of this current determines the resistance of switch capacitor circuits shown in equation 3.18. Switch capacitor circuit has been simulated for clock frequency between 10 kHz - 10 MHz for 20 steps and average current has been calculated for each step. The overall resistance has been calculated and plotted in Matlab which shows a variation of resistance between 1 MΩ to 7 GΩ as shown in figure 3.29.

$$Resistance = (applied\ voltage\ across\ emulator)/(Average\ of\ the\ current\ spikes)$$
$$(3.18)$$

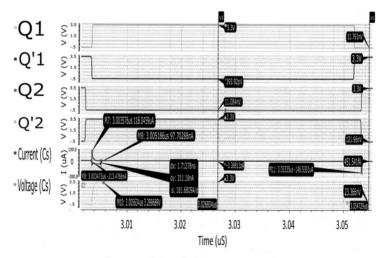

FIGURE 3.28: Switched Capacitor Simulation

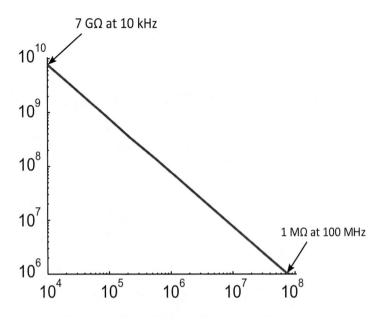

FIGURE 3.29: Switched Capacitor Resistance over Frequency

The fixed capacitor for this switched capacitor has been used to be 8.77 fF. This low capacitance used to achieve high resistance range for the practically implementable frequency range for this technology.

3.2.4 Practical issues with switched capacitor

Switched capacitor (SC) circuit is a promising candidate for memristor emulator where resistance can be tuned by simply changing the clock frequency. This circuit requires less area and has a less complicated circuit. The SC circuit can have a high resistance range but it requires low capacitance and high-frequency variation. To accommodate a resistance range of $1\,M\Omega$ to $7\,G\Omega$, a capacitor of 8.7 fF and frequency range for 10 kHz to 100 MHz has been used. This low capacitance has a severe issue in measurement as the signal is very much susceptible to noise.

The measurement shows a high level of noise compared to the peak current as shown in figure 3.30 which makes measurement values to be different from simulation.

This issues can be solved with a high precision measurement system or high value of capacitor which will eventually lead to lower resistance.

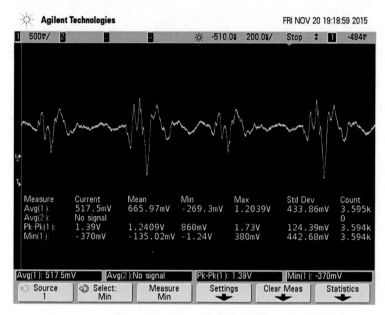

FIGURE 3.30: Switch Cap Measurement

3.3   Resistive switching versus capacitive switching comparison

On-chip resistive and capacitive switching approach has been used to implement a memristive function. Both approaches have their own pros and cons as shown in table 3.3. The switch capacitor approach has high resistance range and low area but it is highly susceptible to noise, switched resistive approach requires large area but its less susceptible to noise and has robust performance.

Parameter	Resistive Switching, section 3.1.1	Switched Capacitor, section 3.2
Area	Much more area required than switch capacitor	Less area required on silicon
Power	More power	Less power
Noise	Less susceptible to noise	More susceptible to noise
Clock frequency	Single clock	Wide range of clock required
Resistance range	Low	High

TABLE 3.3: Switch capacitor versus resistive switching

3.4   Integrate and fire neuron

An analog Integrate and fire neuron circuit has been designed to emulate the principal behavior of biological neurons which can be used for implementing electronic neural systems.

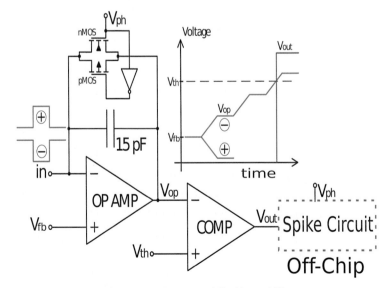

FIGURE 3.31: Integrate and Fire Neuron [62]

The neuron circuit works in two different phases, integration and spiking phase. During the integration phase, it uses pin **in** as an analog input at which it integrates receiving current pulses. The integrator is formed by the operational amplifier with negative capacitive feedback as shown in figure 3.31. When the integrated charge reaches a threshold the circuit triggers the spike generation circuit off-chip. Therefore a comparator is used to generate a trigger voltage whose negative input pin is connected to the output of the **OP AMP**. If the node voltage **Vop** exceeds the adjustable threshold voltage **Vth** then pin **Vout** goes "low" and signals a fire event. At the same moment, a "high" (3.3 V) on the digital input **Vph** must signal the circuit to change from integration into the spiking phase in which the integrator works like a voltage follower. With a "high" on **Vph** the transmission gate is "ON" and connects the output and input of the opamp. The neuron circuit is then not integrating because the integrator is inactive thus making it impossible to generate a new spike. Pin **Vfb** will be used as an analog input to feedback voltage spikes to pin **in** during the spiking phase and to provide the resting potential **Vfb** to that input during integration.

When the spike is over, a "LOW" on **Vph** (0 V) will bring the circuit back to the integration state. The analog voltage input pin **Vth** has been used to adjust the spiking threshold. The corresponding charge threshold can be calculated as follows:

$$Q\_th = 15pF * (V\_th - V\_fb)$$

The threshold voltage has to be larger than the resting potential **Vfb** or the circuit would continuously trigger a spike. Current pulses delivering net positive charges at the input will act inhibitory to the neuron and pulses with net negative charges will act excitatory.

3.5 Integrate and fire neuron measurement

The integration and firing time of neuron for a constant input condition was measured as shown in figure 3.32. The input condition was set as, resting potential(**Vfb**) at 1 V, threshold at 2.7 V while a negative current of 1 nA at input.

In order to change between spiking phase and integration phase, a pulse train was applied to V_{ph}.

A constant input current makes a linear change in V_{op} which when crosses the threshold, the comparator output change from LOW to HIGH as shown in figure 3.32.

The integration time has been measured and $\mathbf{V_{ph}}$ (blue signal) has been displayed which indicates the beginning of integration, up to the moment $\mathbf{V_{out}}$ changes from LOW to HIGH (pink signal). The integration time for neuron is

$$t = \frac{C \cdot (V_{th} - V_{fb})}{I_{in}} \tag{3.19}$$

where C is the feedback cap, V_{th} is threshold voltage, V_{fb} is resting potential and I_{in} is input current.

With the applied input signals and a $15\,\mathrm{pF}$ feedback capacitance, theoretical integration time found by equation 3.19 yields is $25.5\,\mathrm{ms}$. The integration time measured on the oscilloscope (figure 3.32) has been found to be $25.3\,\mathrm{ms}$, which matches theoretical calculation with good accuracy.

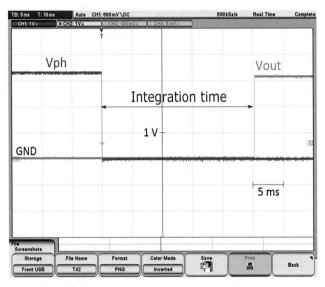

FIGURE 3.32: Integrate and Fire Neuron Measurement [62]

3.6 ASIC Floorplanning

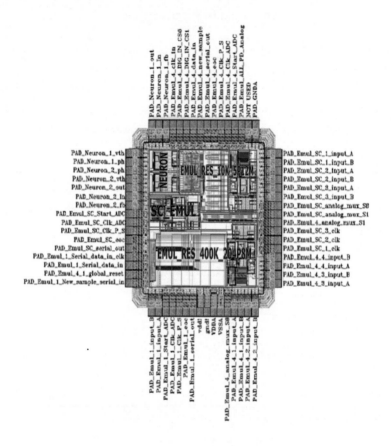

FIGURE 3.33: Emulator ASIC Floorplan with pin details

Nine emulators and 2 neuron circuit have been designed and its layout has been done for final tapeout. The first step of tapeout includes floorplan of top-level ASIC. Neuron designed in this chip is very sensitive to external noise and has to be protected from substrate coupling also. This block has been placed on an extreme top left corner of the chip. All blocks have been placed as per the plan as shown in figure 3.33. All the free space on silicon has been filled with 200 pF of decoupling capacitor.

3.7 Conclusion

Externally programmable memristor emulator ASIC has been designed and tested which allows implementing memristive behavior in a wide voltage and frequency range. The demonstrator has the freedom to choose any external programming tool like MATLAB or FPGA to implement memristive functions. An ion drift based memristive function has been implemented with FPGA which shows the memristive behavior (pinched hysteresis current-voltage relation) on hardware. The measurement shows an impact of operating frequency on memristor behavior as the memristor has higher hysteresis at a lower frequency and memristor acts like a resistor at high frequency. The demonstrator was realized in a 350 nm CMOS technology which offers a very good cost-performance ratio. The circuitry could be easily scaled to smaller dimensions by using e.g. 65nm CMOS technology and adjusted to 1 V operation.

Chapter 4

Neuromorphic pattern recognition

Biologically inspired neuromorphic circuits require a large array of memristor and neurons to perform functions of brain-like pattern recognition, unsupervised learning or cognitive task. Even after the tremendous success of digital computers, it is not able to mimic biological systems due to huge power and device overhead. A biological brain has the power of adaptation and it keeps changing with environmental inputs. This level of complexity and performance of the brain has taken millions of years of evolution. The human brain has a high dense synapse-neuron structure (10^{11} neurons and 10^{15} synapses). Even after a extensive research in neuroscience, only a very basic functionality of brain has been understood. In order to understand the complex structure of the brain, it is important to understand the properties of synapses, neurons, synapse-neuron relationship and basic learning behavior. Basic known functionalities of synapse used in neuromorphic circuits are Long-term potentiation (LTP), Long-term depression (LTD) and synapses plasticity. In this work, first these functionalities have been tested on emulator and with the memristor spice model of $Ti0_2$ based device, pattern recognition has been simulated in LTspice.

4.1 Long-Term Potentiation (LTP) and Long-Term Depression (LTD)

4.1.1 Long-term potentiation

Long-term potentiation ([78] - [79]) explains the phenomenon of synapses strengthening with burst of high frequency stimulation called tetanus in biophysical parlance. The phenomenon of LTD was first proven by Timothy Vivian Pelham Bliss and Terje Lomo in 1973 [80]. His work has provided a neural explanation for learning and memory in the human mind. The memristor is supposed to be the replica of the synapses in neuromorphic engineering and shows LTP characteristic similar to real synapses in the human brain.

The LTP behavior of synapse has been tested on memristor emulator by applying series of potentiation pulses and measuring conductance. Rectangular pulses of 3 V and pulse width of 500 μs have been applied and emulator shows the increase in conductance like real synapse as shown in figure 4.1.

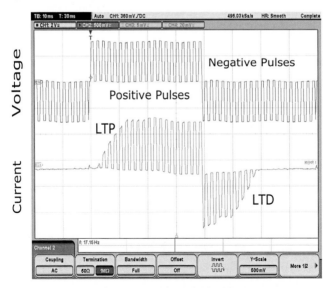

FIGURE 4.1: Screenshot of LTP and LTD measurement [61]

4.1.2 Long term depression

LTD is the inverse phenomenon of LTP (4.1.1) where negative polarity of depression pulses make synapse more resistive and eventually stops at highest resistance [81], [82], [83]. LTD has been measured by applying negative pulses and measuring the resistance across the emulator. As shown in figure 4.1 consecutive negative pulses shows an increase in resistance which leads to a reduced current through memristor.

4.2 Synaptic plasticity

Synaptic plasticity is the gradual change in conductance and adjustment in the synaptic conductance between two neurons based on the applied voltage across the synapses. Synaptic plasticity is the core phenomenon of spike time dependent plasticity (STDP)[84],[85] which is the basic mechanism for neural learning [86]. The memristor model as explained in section 3.1.4 has been tested with 50 potentiation and depression pulses of amplitude $1\ V$ and pulse widths $500\ \mu s$ and $900\ \mu s$. Both pulses show a different rate of change of conductance. $900\ \mu s$ pulses have a higher rate of change of conductance compared to $500\ \mu s$ as shown in figure 4.2.

4.3 Memristive pattern recognition

Artificial neural network (ANN) has been found to be an effective method for works which requires large data handling like pattern recognition, clustering, classification etc. Software based ANNs are commercially available in market but are insufficient for increasing complex networks. An alternative approach is hardware neural network (HNN) which uses analog and digital circuits to implement neurons and synapses [87], [88], [89], [90], [91], [92] and have speed advantage over software based approach [93], [94], [95].

Memristive pattern recognition utilizes integrate & fire neuron and memristor as fundamental element. Architecture from [41] has been used and simulated by Wolf Lukas Hellweg in his master thesis [96]. Memristive pattern recognition system shown in figure 4.3 has a matrix of memristors, input and output neurons.

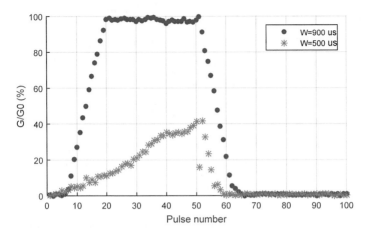

FIGURE 4.2: Synaptic plasticity measurement. It shows that for the same number of potentiation spikes across memristor, change is conductance depends on the pulse width of spike. Spikes of higher pulse (900 μs) width shows greater change in conductance than spikes of lower pulse width (500 μs) [61]

The architecture has **n input** and **m output** neurons. Each output neuron is connected with each input neuron through a memristor. Memristor **00** represents the memristor across input neuron **1** and output neuron **1**.

An input image of 12 pixels (4 row and 3 column) has been used for unsupervised learning. Four such images (figure 4.4) is presented in sequence at the input and with time architecture adapts to particular images and each output neuron learns a particular pattern. Input pixel in encoded as positive (0.6 V) and negative (-0.6 V) for the bright and dark pixel as shown in figure 4.3. During learning when a particular neuron fires, it generates a bipolar spike of positive and negative voltage. When these output spikes interact with input image pixels via memristor, it causes potentiation or depression in memristor which is the core of memristive learning and called Spike timing dependent plasticity (STDP). When an output neuron fires, a spike appears at the output as well as the input of output neurons. The generated spike appears at the negative terminal of all memristors connected with the output neuron. Set of memristors which received bright pixel at input interacts with these spikes and cause a potentiation in memristor making it more conductive than other memristors. As shown in figure 4.4, when output neuron **1** fires, memristor **00** receives a white pixel at input and memristor **n0** receives dark pixel at input which causes potentiation in memristor **00** and depression in **n0**.

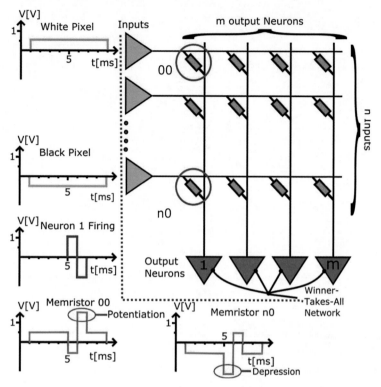

FIGURE 4.3: Network architecture and mechanism [97]

In addition to memristor and neurons, three important elements for learning are competition, homeostasis, and variability.

4.4 Competition, homeostasis and variability

Competition is a phenomenon in the neural system where each neuron competes for firing and resets all other competing as soon as any one of the output neurons reaches a threshold. As shown in figure 4.3 each output neuron is connected with all other output neuron in a winner-take-all (WTA) topology [98], [99]. When the input spike appears at input neuron, it appears at the positive terminal of all memristors and allows current through memristor to integrate into each output neuron. The output neuron which

64

reaches threshold first resets all others neuron to zero potential. This competitive environment enables each neuron to be receptive to an individual image. The neuron which fires first has high conductance towards bright pixel and low conductance toward dark pixel and only this neuron gets receptive toward the current image.

Competition has a disadvantage that the first firing neuron can get receptive to more than one image due to an overlapping pixel in different images. This may make the system unstable causing a single neuron to learn all image and the majority of the neurons may not take part in learning. This issue is solved by homeostasis in biological brain.

Homeostasis is the adjustment of neuron threshold to maintain equilibrium. Neuron which firest faster than the target rate is panelised by increasing its threshold and vice versa. This allows each neuron to adjust its threshold to maintain stability and every neuron get opportunity to fire [100], [101], [46], [102]. Homeostasis regulates the competition based on the firing rate of a neuron according to the equation as shown in equation 4.1.

$$\frac{dV_{th}}{dt} = \gamma(R_{avg} - R_{target}) \qquad (4.1)$$

where, R_{avg} is average firing rate of neuron, R_{target} is target firing rate, γ is the gain factor and $\frac{dV_{th}}{dt}$ is the rate of change of threshold. This homeostasis control was first proposed in [41] and explained in detail in [103]. According to this scheme when an output neuron fires faster than R_{target}, V_{th} increases and vice versa (equation 4.1).

Variability is the variation in the initial state of memristors. In neuromorphic learning it introduces the initial point for neuron differentiation. This variability is needed for pattern recognition and additional variability is introduced in input coding for better learning. This has been explained in detail in [97], [62] and [96].

4.5   Simulation results

The output neuron develops its receptive field towards a particular image during learning as shown in figure 4.5. This learning is the cumulative effect of competition, homeostasis, and variability. The simulation result shows that two patterns have developed their receptive field for image 1 and 2 in 400 ms. As shown in figure 4.5, in the initial stage of learning pattern 3 is receptive to all neuron 2, 3 and 4 which is undesired. But due to homeostasis and variability, each neuron relocates its receptive field towards unrecognized patterns and slowly after 200 cycles, each neuron learns a particular image. The spice model for the memristive device has been developed by the Kiel university ([14]) and modified by Mr. Hellweg in his master thesis work to fit for neuron circuit described in section 3.4. The architecture is explained in detail in [41].

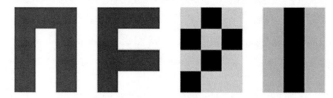

FIGURE 4.4: Basic Patterns to be learned by the ANN [97]

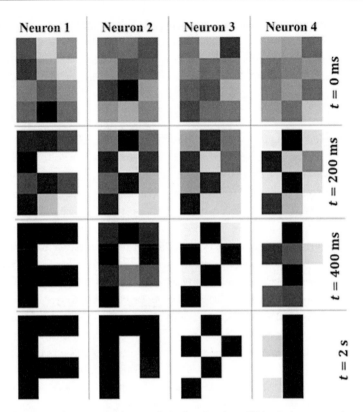

FIGURE 4.5: Receptive development during learning phase. White pixels correspond to a memristor state of w=1 and black to w=0. Each neuron get receptive to a particular image after 2 seconds of learnng. Neuron 1 gets receptive to 2nd input image, neuron 2 to 1st, neuron 3 to 3rd and neuron 4 to 4th input image [97]

4.6 Memristor based neuromorphic versus digital computer approach

Neural Networks can be implemented in digital computers (Software-based ANN) as well as in hardware (Memristor-based HNN) as shown in figure 4.7. Architecture shown in figure 4.7 has 12 inputs and 4 outputs. It shows the architecture for learning four

67

different patterns by four output neurons. Similar architecture with the same input/output characteristic has been used for comparison between neuromorphic and VNA approach. Considering the input coding and homeostasis takes the same resource in both approaches, HNN and ANN have been compared here regarding the technical challenges in implementing neural networks in digital computers and device count in both approaches.

Software-based ANN utilizes a digital computer to store program and data in non-volatile memory and the neural network is executed in a microprocessor (figure 4.7). As shown in software-based ANN, each neuron is connected with 12 input neurons via synaptic weight which can be modified by the processor and the processed data can be stored in memory. $W(1,1)$ is the synaptic weight between input neuron 1 and output neuron 1. Similarly, each output neuron is connected with each input neuron (input pixel) by a synaptic weight. Implementing the same learning algorithm as explained in section 4.3 and 4.4 in software, requires adjustment of synaptic weight depending on spike timing between presynaptic neuron and post-synaptic neuron firing. Weight adjustment in software is a cumbersome process unlike memristor based HNN. Each synaptic weight is accessed, processed and updated in memory after each firing event of any neuron. Each time an output neuron fires, all the synaptic weight connected with that neuron has to be accessed by CPU via high bandwidth bus. The amount of data to be transferred depends on the resolution of the image. In this example, each weight has a bit depth of 8 bit which is standard bit depth of the modern-day display. So, each firing event transfers a digital data of $12 \times 8 = 96 \; bits$. Considering the human recognition time (100 ms) as a reference, to get a perception of human level, $960 \; \frac{bits}{second}$ of data has to be transferred between memory and processing. This data rate grows rapidly when the computer has to recognize a high-resolution image. Digital computers have challenges in this high volume of data communication due to memory wall (decided by DRAM speed) and communication bottleneck. Figure 4.6 shows state of art memory to processor communication speed and rise in image quality with time. A human mind can perceive 10 images per second, so each high definition resolution is first converted in a total number of bit (including the color depth) and multiplied by 10 to get bit rate in Gbps. This is the rate at which the digital computer needs to process data for reliable learning as per human level. However, state of the art communication speed between memory and processor is 11 Gbps. Any image above this resolution requires different algorithm to be recognized by ANN. Figure 4.6 shows that even now to recognize state of the art 8K UHD image by a neural network will require higher communication speed than state of the art communication speed of 11 Gbps. It shows that digital computer has reached the saturation level to learn high-resolution image of the present world.

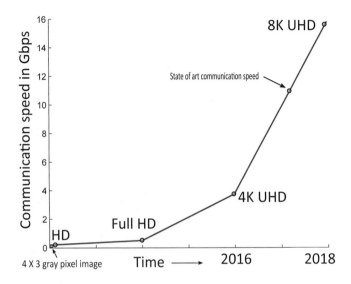

FIGURE 4.6: Digital computer (VNA) communication speed versus rise in image resolution. Each new generation of high resolution image requires high speed communication between DRAM and processor. Embedded DRAM on microprocessor may be an option which is costly and needs an architecture change. Neuromorphic processing are capable of handing such application by extending the HNN architecture in array (figure 4.7)

Memristor based neuromorphic processing does not require weight adjustment by different processing unit as shown in figure 4.7. The processing and storage takes place at same location in memristor. This provides freedom to train any high resolution of image on memristor based HNN. HNN shown in figure 4.7 has memristor crossbar array and each output neuron is connected with each input neuron via a memristor. Each input image is 4X3 pixel, so for learning four different image 48 memristors are required. M(1,1) in figure 4.7 shows the memristor connected between first input neuron and first output neuron. Similarly, M(12,4) show the memristor connected between 12th input pixel neuron and 4th output neuron. This approach as explained in 4.3 shows successful learning of 4 images. This architecture can be expanded up to 8K UHD resolution and pattern recognition of ultra high resolution image can be performed which is a challenge for modern day digital computers. Moreover, memristors utilises less number of devices compared to digital memory. State of the art bit depth of each pixel is 10 bit and 10 memory cells are required to store each pixel. Even with flash memory which can store 3 bit in a cell, at least 3 bit will be required. Memristor has contineous memory and single memristor can have 10 bit of resolution which shows at least 3 times less device required in memristor based approach.

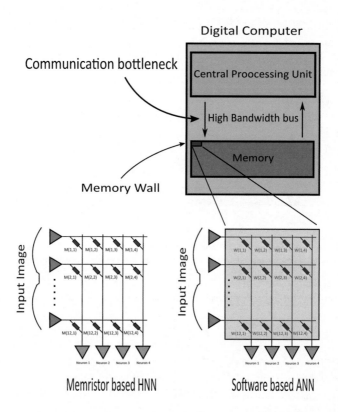

FIGURE 4.7: HNN has memristor array capable of processing and storing data at the same location. The memristive array can be fabricated in cross-bar structure and can extend to any size without much change in hardware architecture to accommodate a high volume of data processing. ANN has a separate memory which requires high bandwidth communication which limits its application in high volume data applications and requires hardware modifications like embedded DRAM on processor or multi-core processors.

4.7 Conclusion

LTP, LTD and synaptic plasticity are the key requirements of neuromorphic learning and have been tested on the demonstrator. An integrate and fire neuron circuit has been implemented and tested which has similarity to biological neurons. A neural network architecture has been simulated on LTspice based on implemented memristor and neuron model. The simulation results show that the circuit was able to learn and recognize a 3 X 4 image pixel image. This paves the way towards real-time large image recognition on hardware with memristor and CMOS circuits. Pattern recognition can be done more efficiently with the neuromorphic approach i. e. fewer hardware elements are required than in a CMOS circuitry with the same capability. Moreover, the neuromorphic approach has localized processing and analog storage which allows handling huge data without bandwidth limitation.

Chapter 5

Neuron ASIC

Neuron ASIC is CMOS circuit implementation of a biological neuron. This circuit is implemented in the AMS350nm process and 3.3V supply. The neuron circuit has on-chip Integrate & Fire circuit, spike generator, and homeostasis control. Integrate and fire circuit has an operational amplifier in integration mode with a comparator to fire when the integrated current reaches a particular threshold. This is widely used neuron circuit for neuromorphic applications [104], [105], [106]. Spike generator circuit is externally controlled and it has positive and negative spikes with respect to virtual ground. The width of the spike generator is externally controlled by changing the clock frequency of the spike generator. The chip contains 15 neurons on-chip with each neuron having a driving capacity of 84 pF capacitance.

5.1   Neuron floorplan

Neuron circuit consists of integrator, comparator, biasing circuit and digital control circuit as shown in figure 5.1. The integrator has been placed at the input of chip close to pad to avoid any parasitic effect and reduce input pin routing. The input routing pin has been fully shielded co-axially to avoid any interference of any close by signals with input spike. Integrator Output pin (INT_OUT) has been placed close to the comparator with the least possible routing to avoid any noise coupling. This signal is very sensitive to noise and can trigger comparator due to noise. This node is also co-axially shielded and connected to the negative terminal of the comparator. Both terminals of the comparator have high input impedance. Neuron threshold pin is off-chip able to adjust any threshold value. When the comparator triggers, a trigger signal is sent to the digital circuit which generates a spike. The spike width is externally controlled by the frequency of the clock.

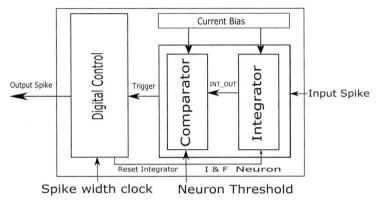

FIGURE 5.1: Neuron circuit floorplan of ASIC

5.2 Integrate-and-fire neuron operation

Neuron operation can be divided into two phases: Integration phase and Spiking phase (5.2). During integrator phase neuron integrates all incoming currents and when integrator output voltage (V_{out}) reaches a particular threshold it enters spiking phase and generates a spike for other neurons. The phase is decided by the V_{ph} signal which when high, turns on transmission gate and discharges integration capacitor. During this time spike generator circuit is activated and output of neuron sends a spike to output.

Integration phase

During integration phase, **V_fb** signal is at 2.5 V as explained in section 5.3.3. Due to input regulation of operational amplifier, the **In** terminal of op-amp also remains at 2.5 V. When input spike appears at "input spike", it allows current to flow through capacitance as transmission gate is in high resistance state (off state). Integration capacitor integrates incoming current and output voltage (V_{op}) starts reducing. The rate of change of **Vop** depends on the input current and capacitor by the equation

$$\frac{dv}{dt} = \frac{I}{C} \tag{5.1}$$

When V_op reaches the threshold voltage of the comparator (V_{th}), the comparator changes its polarity from low to high and digital circuit is activated which triggers neuron enters into spiking phase.

Spiking phase

When comparator output changes its polarity from zero to high (5 V), digital spiking circuits gets triggered. It generates high spike (5 V) for 16 clock cycles of input **spike clock** then low (0 V) for next 16 **Spike Clock** as explained in section 5.3.3. After 32 clock pulses the spike generator circuit again settles at default voltage level of 2.5 V. Since, spike out signal is connected to positive terminal of OP-AMP, the spike appears at **Vfb** as well as (**In**) terminal. In spiking phase, digital circuit generates high **Vph** for 32 clock cycles. The transmission gate is in short state during this period and spike signal appears at (V_{op}) as shown in simulation result 5.3.

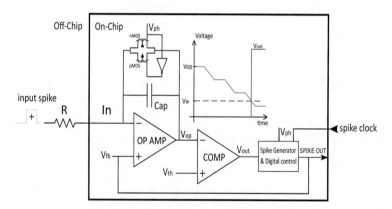

FIGURE 5.2: Neuron circuit with fixed resistor [63]

During the spiking phase, a neuron cannot spike as the integration is stopped. This is analogous to the refractory period of a biological neuron.

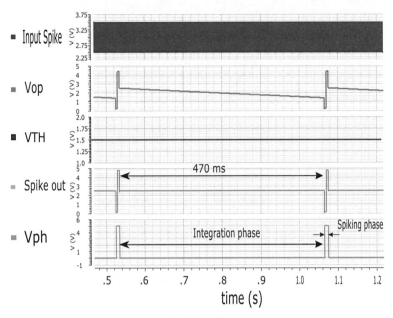

FIGURE 5.3: Simulation of neuron circuit

5.3 Neuron circuit blocks

Neuron circuit includes 3 important design blocks: Operation amplifier, comparator and spiking circuit.

5.3.1 Operational amplifier

A two-stage CMOS Operational Amplifier (OP-AMP) has been designed as shown in figure 5.4. Each memristor has parasitic capacitor up to 7 pF. 12 such memristors had to be driven by each op-amp for recognizing a 4 X 3-pixel pattern. So, each op-amp has a high driving capacity of 84 pF. Considering the high range of memristor to be connected at the input of the op-amp, its input has to be high, so MOSFET has been used at the input stage. Table 5.1 shows the specification of the designed OP-AMP. Operational amplifier design required following design considerations.

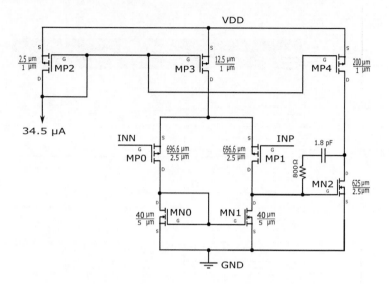

FIGURE 5.4: Two stage operational amplifier designed in AMS 350 nm process.

PARAMETER	VALUE	UNIT
Quiescent Current	2.94	mA
Power Supply	5	V
GBW	7.69	MHz
Slew Rate	5	$V/\mu S$
Open loop gain	86	dB
Output Load	84	pF
Output Integrated Noise	84	μV

TABLE 5.1: Operational Amplifier Specification

Open loop gain and gain bandwidth

Operational amplifier open loop gain (A_0), Gain bandwidth (GBW) remains the first important specification in design. For an integrator to work like ideal over a frequency range of 1 Hz to 700 kHz (operating frequency range of DBMD learning), open loop gain and GBW must fulfill the condition shown in equation 5.2 and explained in detail in [107].

$$10 \cdot \frac{w_I}{A_0} < f < \frac{GBW}{10} \qquad (5.2)$$

Where, w_I is integrator frequency and $\frac{1}{w_I}$ is the integrator time constant which depends on input resistance and feedback capacitor of integrator (equation 5.3). GBW is unity gain bandwidth of opamp.

$$w_I = \frac{1}{RC} \qquad (5.3)$$

Considering minimum ON resistance of memristor to be 50 MΩ (based on measurement results) and integration cap of $11.4~pF$, according to equation 5.2, A_0 has to be greater than 17543. The operational amplifier has been designed with relaxed specification with opamp loop to be 86 dB (approx. 20000) for better accuracy.

According to equation 5.2, GBW of 7 MHz is required for OP-AMP to operate as ideal integrator. GBW of 7.69 MHz has been used in this design for better performance.

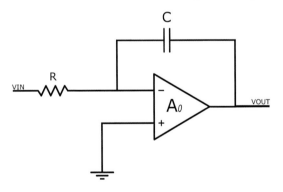

FIGURE 5.5: Inverting operational amplifier with feedback capacitor as an integrator

Input devices : NMOS versus PMOS

NMOS device shows better current efficiency than PMOS as shown in figure 5.7, but PMOS device has been used here because PMOS devices have lower flicker noise than

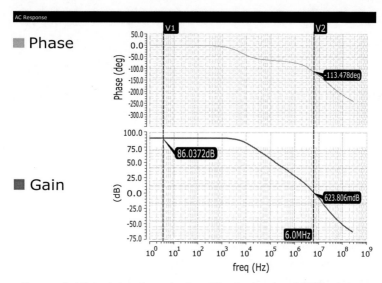

FIGURE 5.6: AC simulation of operational amplifier showing A_0 and GBW and phase margin of 67 °

NMOS and PMOS devices shows better matching than NMOS. Flicker noise is inversely proportional to area (width and length) as shown in equation 5.4.

$$V_f^2 = \frac{K_F}{C_{ox}^2 \cdot W \cdot L \cdot f} \tag{5.4}$$

K_F for NMOS is higher than PMOS, So NMOS contribute higher flicker noise than PMOS in AMS 350 nm process.

Channel Length

Channel length decision is the first step in deciding device other parameters. The channel length of a device has an impact on threshold, gain and flicker noise. Flicker noise does not depend directly on the length of the device but on gate area of the device, so this is not deciding factor for channel length. Channel length has an impact on the gain of the device but it is also not the only parameter for gain. Gain is dependent on the ratio of $\frac{g_m}{g_d}$ so length cannot be deciding factor for length decision.

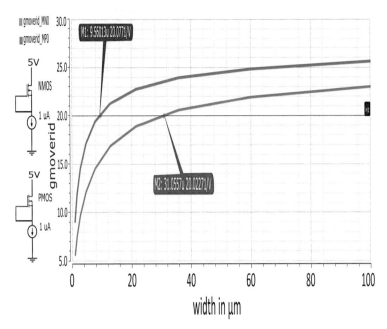

FIGURE 5.7: Transconductance (gm) over drain current (id) for NMOS and PMOS versus device width for fixed input current (1 μA). X-axis gives channel width which determines transconductance of device.

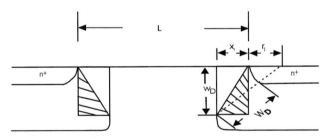

FIGURE 5.8: Mosfet device cross-section [108]

Change in the threshold of a device is inversely proportional to channel length explained by equation 5.5.

$$\delta V_{th} = -\frac{q N_B w_D r_j}{C_{ox} L} \tag{5.5}$$

where N_B is bulk doping concentration, w_D and r_j are physical dimensions shown in figure 5.8 , C_{ox} is the oxide capacitor per unit area and L is the channel length. Derivation of equation 5.5 has been explained in [108]. For a fixed current of $1\ \mu A$ and width of $100\ \mu m$, length has been swept over a range from $0.4\ \mu m$ to $50\ \mu m$ and Vth has been simulated. As shown in simulation result in figure 5.9, devices shows stable Vth at 2.5 μm. In this design, channel length of 2.5 μm has been used for integrator and comparator design.

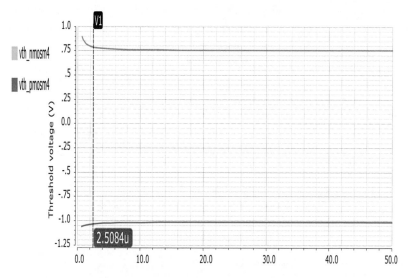

FIGURE 5.9: Variation in NMOS and PMOS device threshold voltage with variation in length for fixed width of 100 μm.

Slew rate

Slew rate is the rate at which operational amplifier can transit from 0 to 5 V. This depends on current I_{MP3} and compensation capacitor (C_c) as well as I_{MP4} and load capacitor (C_L) as shown in equation 5.6 and 5.7. The slew rate of operational amplifier has been decided based on the input pulse and maximum swing of the operational amplifier. The input spike may appear at the input at a rate of maximum 700 kHz, time period 2 μs, the pulse width of 1 μs. For proper operation the output of operational amplifier has to change from 0 to 5 volt in 1 μs, making slew rate to be $\frac{5\ V}{1\ \mu s}$.

$$\frac{I_{MP3}}{C_c} = SR \qquad (5.6)$$

$$\frac{I_{MP4}}{C_L} = SR \qquad (5.7)$$

Limiting slew rate is the one which is slower. Equation 5.6 provides a slew rate of $\frac{6.9V}{1us}$ which is the deciding slew rate. This slew rate is sufficient for this application as the normal operating frequency will be 1 kHz. Slew rate has been simulated for all corners and found to be acceptable for this application.

Input gate leakage

Input gate leakage is an important consideration for this opamp as the input current is in pA range and gate leakage must be in fA range to avoid unwanted integration and firing. The simulated opamp shows leakage current in the range of $50fA$ which negligible for this application and shows no detrimental impact on charging time of integrator.

Integrator ideal behavior and stability

Integrator stability has been tested and found to have a phase margin of 67 ° as shown in figure 5.6. For an integrator to work like ideal between the memristor range of $100\,\text{M}\Omega$ and $200\,\text{M}\Omega$, equation 5.2 has to be satisfied and integrator has to be stable in full range of operating frequency [107].

AC simulation for the integrator circuit (figure 5.5) has been done. The simulation shows the phase margin of 90 ° over the frequency range between 0.1 Hz and 700 kHz. The maximum operating frequency (f_{max}) is limited by GBW of the opamp and minimum frequency (f_{min}) is limited by open loop gain (A_0), feedback capacitor (C) and input resistance (R).

Noise analysis

Noise is the unwanted signal which limits the minimum signal level a circuit can process with acceptable quality. Noise is a random signal and its voltage level cannot be predicted at any time. The random nature of noise is analyzed by the average power of noise which can be modeled by equation 5.8.

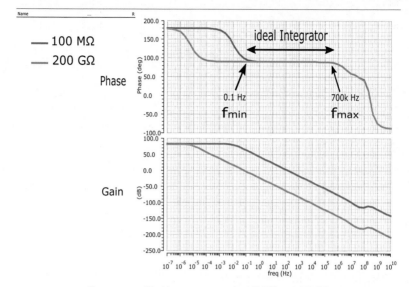

FIGURE 5.10: Ideal integrator range for 100 MΩ and 200 GΩ

$$P_{avg} = \lim_{T \to \infty} \frac{1}{T} \int_a^b v^2(t)dt \tag{5.8}$$

where $v(t)$ is random noise voltage and P_{avg} is the average noise power.

Analog circuit has most dominantly thermal and flicker noise which is generated internally and needs to be taken into consideration.

Thermal Noise

Thermal noise in a circuit is due to the random motion of electrons due to thermal agitation in an atom. Thermal noise is dependent on temperature and a resistor has spectral noise density [109] shown in equation 5.9.

$$S_v = 4kTR \tag{5.9}$$

where $k = 1.38 \times 10^{-23}$ J/K is the Boltzmann constant and S_v is noise power per unit frequency (V^2/Hz or $\overline{V_n^2}$). Total noise of a circuit can be obtained by integrating spectral noise density over the bandwidth of the circuit [109]. Any RC circuit can be modeled by a noiseless resistor and a noise voltage source as shown in figure 5.11. The total integrated noise of an RC circuit is shown by equation 5.10 which is total integrated noise over the bandwidth of the circuit. Total noise is inversely proportional to capacitor (C), increasing C will reduce total noise but reduce the bandwidth of the circuit. It is a challenge in an analog circuit to design high bandwidth circuit with less noise.

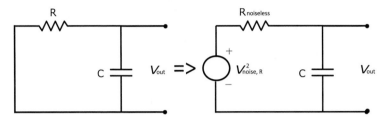

FIGURE 5.11: RC noise

$$V_{noise,R}^2 = \frac{kT}{C} \tag{5.10}$$

Thermal noise in MOSFET device is most dominantly generated in the channel. For long channel MOSFET device operating in saturation, the channel noise can be modeled by a current source as shown in figure 5.12.

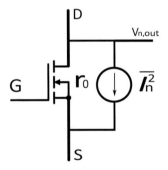

FIGURE 5.12: MOSFET output noise current

Spectral noise current density of MOSFET in saturation region depends on transconductance (g_m) of device and on temperature [109] as shown in equation 5.11. Output impedance (r_0) of MOSFET introduces an output noise voltage ($V_{n,out}$) as shown in figure 5.12. Spectral output noise power density of MOSFET can be modeled by equation 5.12. As seen in equation 5.12, $V_{n,out}^2$ is dependent on output gain and it indicates that as

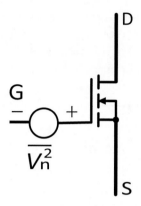

FIGURE 5.13: MOSFET input noise voltage

gain of circuit increases, the circuit gets more noisy but this is not true as the gain amplifies the input signal in same ratio. So it is more accurate to analyze the input referred noise of the circuit which is output noise divided by gain as shown in equation 5.13.

$$I_n^2 = 4 \cdot k \cdot T \cdot \gamma \cdot g_m \tag{5.11}$$

$$V_{n,out}^2 = I_n^2 \cdot r_0^2 \tag{5.12}$$

$$V_{n,in}^2 = \frac{V_{n,out}^2}{Gain} = \frac{I_n^2}{g_m^2} \tag{5.13}$$

where g_m is the transconductance of a device in saturation, k is Boltzmann constant, T is temperature and γ is a technology dependent constant and it is approximately equal to $\frac{2}{3}$.

Flicker Noise

Flicker noise appears in the channel current due to dangling bonds at the silicon and silicon oxide interface. These dangling bonds trap and release electrons randomly and causes flicker noise in the channel. The flicker noise in not only caused by the trapping phenomenon but some other mechanism are also involved which is explained in [110].

Flicker noise unlike thermal noise is highly dependent on the technology and is modeled by equation 5.14 [110].

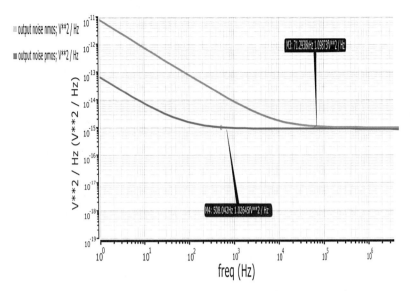

FIGURE 5.14: Spectral noise density of NMOS versus PMOS for fixed transconductance and current efficiency($\frac{g_m}{I_d}$)

$$V_{n,out}^2 = \frac{K_f}{C_{ox}WLf} \qquad (5.14)$$

Where K_f is technology dependent constant and is different for NMOS and PMOS. In AMS350 nm process, for 5 V devices, K_f for NMOS device is $K_f = 9.18 \times 10^{-26} \ V^2F$ and for PMOS device it is $K_f = 3.625 \times 10^{-26} \ V^2F$. NMOS device has approximately 3 times greater coefficient than PMOS device due to which NMOS devices contribute 3 times more flicker noise than PMOS. For low-frequency operation where flicker noise is a major concern, PMOS devices at the input will introduce less noise than PMOS at input state.

Characteristics of flicker noise are analyzed by calculation flicker corner frequency of circuit which is the frequency at which thermal noise and flicker noise spectral power density are same. The calculated flicker corner frequency (f_C) is shown in equation 5.15.

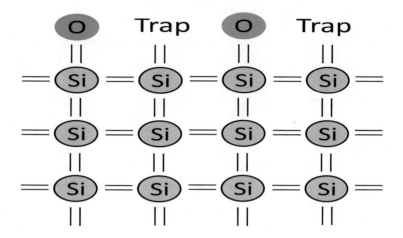

FIGURE 5.15: Flicker noise due to dangling bond at silicon and silicon oxide interface.

$$f_C = \frac{3K_f g_m}{8C_{ox}WLfkT} \tag{5.15}$$

To make a fare comparison of devices, 5V NMOS and PMOS devices has been simulated for same $\frac{g_m}{I_d}$ and input referred noise has been plotted and shown in figure 5.14. For 1 μA of current $\frac{g_m}{I_d}$ of 20 has been used which keeps devices in saturation. The width of both device for keeping same $\frac{g_m}{I_d}$ has been calculated by the simulation shown in figure 5.7 which shows for 1 μA of current NMOS ($\frac{9.5\mu}{2.5\mu}$) and PMOS ($\frac{31\mu}{2.5\mu}$) is required to maintain same $\frac{g_m}{I_d}$ of 20. PMOS devices has less flicker noise and lower flicker corner frequency because of bigger width and flicker coefficient K_f as shown in figure 5.14.

Noise in double barrier memristor

The noise of a memristor device is an important consideration in designing the neuron input stage as the unwanted high noise can charge the integrator and generate unwanted firing. Noise model of DBMD has not been characterized yet but based on the device physical description in [51], an approximate noise model can be derived which gives a primary noise characteristic for the device.

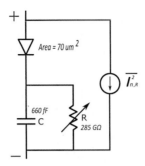

FIGURE 5.16: Simplified noise model of DBMD.

The simplified circuit model shown in figure 5.16 has been simulated for minimum capacitor mentioned in [51]. The capacitor C_M is the total capacitance of interface barrier (C_I) and tunnel barrier (C_T). For maximum total integrated noise, capacitance should be minimum [109]. The memristor device area can vary from 70 μm^2 to 2300 μm^2 and minimum area will produce the minimum capacitance. Based on measured result, $C_T = 2.07 \times 10^{-14} \frac{F}{\mu m^2}$ and $C_I = 1.74 \times 10^{-14} \frac{F}{\mu m^2}$. The calculated total capacitor for 70 μm^2 is equal to $C_m = 1.74 \times 10^{-14}$ F. Minimum area produces maximum resistance and from the measurent in [51] for 70 μm^2 the expected resistance (R_m) is 2.85×10^{11} Ω. The circuit in figure 5.16 has been simulated for C_m, R_m and a diode from AMS 350nm process to calculated the approximate total integrated noise produced. Noise analysis has been done for the simplified model for a bandwidth of 100 MHz (bandwidth of integrator, figure 5.10) resulting in a total integrator RMS noise current of $7.6 \times 10^{-11} A$, which agrees with the measurement result as shown in figure 5.17.

Integrator input and output noise

Noise produced by memristor is also accompanied by the noise produced by the integrator at output and noise produced by comparator at the input. The total integrated noise at the output of the integrator during the integration phase produces total output noise which may create unwanted triggering of the comparator as shown in figure 5.18. The circuit has been simulated for total integrated noise at **VOUT** over the bandwidth of 10 MHz. Total integrated RMS noise voltage is 84 μV. This neuron will receive noise from all memristors in the array. Since the total neuron has to recognize 4 patterns each having a 4X3 pixel, so the total number of memristor in the array is 48. Total integrated noise at output neuron can be roughly computed by equation 5.16 to be 4 mV (48 * 84 μV = 4 mV).

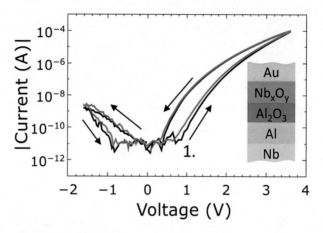

FIGURE 5.17: Two consecutive I-V-curves of the double barrier memristive device. The I-V-curve shows the set process for positive voltages and the reset for negative voltages. Inset: Schematic layer sequence of the double barrier memristive device [63]

$$Total\ noise = number\ of\ memristors \times noise\ contribution\ due\ to\ each\ memristor$$
$$(5.16)$$

This is the simplified approximate calculation for designing the next stage of design and avoid any oscillation at the comparator output. Since the exact model of noise is not still known and that can only be compensated by designing the comparator with higher hysteresis. Integrated output noise of the integrator has 4 mV V_{vrm} and hysteresis of 4 mV V_{vrm} is needed for the reliable firing of the comparator. In this design, a hysteresis of 50 mV has been introduced because comparator fires in only one direction (when integrator output goes below threshold). So hysteresis only increases the threshold by 50 mV and eliminates the chance of comparator oscillation due to noise.

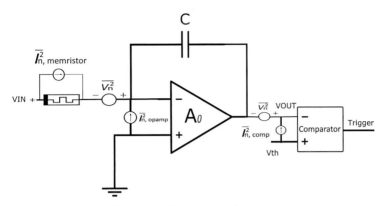

FIGURE 5.18: Integrated noise in neuron

5.3.2 Comparator design

Comparator with internal hysteresis including an output stage has been designed as shown in figure 5.19. In this circuit, there are two feedback paths. The first path is current-series feedback (MN1, MN4) feedback through the common-source node of transistor MP0 and MP1. This is a negative feedback path. The second path is voltage-shunt feedback through the gate-drain connections of transistors MN2 and MN3. This path is positive feedback. When positive feedback is less than negative feedback, over-all feedback is negative and no hysteresis happens. If positive feedback is more than negative feedback, overall feedback is positive and will give rise to hysteresis.

Supply for this comparator is at 5 V and ground at 0 volts. When the gate of MP0 is tied to 2.5 V and input of MP1 much higher than 2.5 V, MP0 is ON and MP1 is off, thus turning on MN1 and MN2 and turning off MN3 and MN4. All of I3 current flows through MN1 and MP0, so V0 (drain of MN3) is low. Low V0 turns off MN2 resulting in high comparator output. At this state, MN2 is attempting to source the following amount of current:

$$I_{MN2} = \frac{(W/L)_{MN2}}{(W/L)_{MN1}} I_{MP3} \qquad (5.17)$$

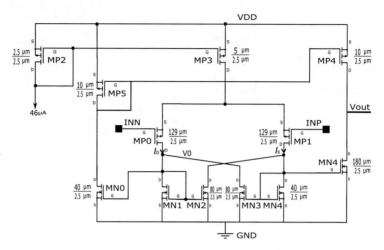

FIGURE 5.19: Comparator with internal hysteresis

As gate voltage of MP1 decreases, some of tail current I3 starts flowing through MN2. This continues till the point when the current in MN2 equals MP1. At this point comparator toggles.

At this point:

$$I_{MN2} = \frac{(W/L)_{MN2}}{(W/L)_{MN1}} I_{MN1} \tag{5.18}$$

$$I_{MP1} = I_{MN2} \tag{5.19}$$

$$I_{MP3} = I_{MP0} + I_{MP1} \tag{5.20}$$

So,

$$I_{MN1} = \frac{I_{MP3}}{1 + \left(\frac{(W/L)_{MN2}}{(W/L)_{MN1}}\right)} = I_{MP0} \tag{5.21}$$

$$I_{MP1} = I_{MP3} - I_{MP0} \tag{5.22}$$

When currents in both MP0 and MP1 is known, V_{GS} of MP0 and MP1 is easy to find by the equation :

$$VGS_{MP1} = \sqrt{\frac{2 * I_{MP1}}{\mu * C_{ox}}} + Vth_{MP1} \tag{5.23}$$

$$VGS_{MP0} = \sqrt{\frac{2 * I_{MP0}}{\mu * C_{ox}}} + Vth_{MP0} \tag{5.24}$$

$$VGS_{MP0} = \sqrt{\frac{2 * I_{MP0}}{\mu * C_{ox}}} + Vth_{MP0} \tag{5.25}$$

$$V_{+,trip_point} = VGS_{MP0} - VGS_{MP1} \tag{5.26}$$

When the threshold is reached, a majority of current flow through MP1 and MN4. Similarly like positive trip point, the negative trip point can be calculated which is also:

$$V_{-,trip_point} = VGS_{MP0} - VGS_{MP1} \tag{5.27}$$

Since these equations do not take channel length modulation into account, width adjustment needs to be done for the devices to achieve hysteresis. With the slight adjustment of devices as per theoretical calculation, the comparator has been designed with the specifications shown in table 5.2.

PARAMETER	CONDITIONS	VALUE	UNIT
Quiescent Current	-40° C to 150° C	276	μA
Power Supply		5	V
Slew Rate		5	V/μs
Hysteresis		50	mV

TABLE 5.2: Operational Amplifier Specification

5.3.3   Spike generator and digital control

The digital module of neuron generates a spike and control signal (V_{ph}) for changing between the integration phase and the spiking phase as shown in figure 5.20. This module has a voltage divider (R = 100 kΩ) connected between V_{dc} (5 V) and Ground. The middle node (SPIKE_OUT) of the divider is connected to the output of a three-state buffer. The functionality of this circuit has been simulated in Fig. 5.21. When the buffer is in high impedance state, the node SPIKE OUT is at its default state and equal to half V_{dc} (2.5 V). When a positive edge is detected on the input V_{out}, the SPIKE OUT output transitions to ground and V_{ph} signal transits to high (5 V). A counter starts and runs for 16 clock cycles (*spiking clock*). After 16 clock cycles (*spiking clock*), a second transition moves this node to V_{dc}. Finally after 32 clock (16 clocks for high and 16 clock for low) cycles SPIKE OUT is set back to its default value of 2.5 V. During this period, V_{ph} is kept at a high value 5 V to keep the transmission gate ON and neuron in spiking phase (Fig. 5.3), (Fig. 5.2).

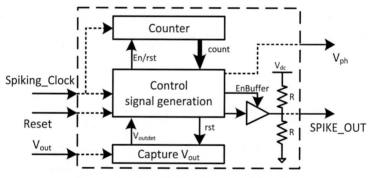

FIGURE 5.20: Spike Generator module [63]

5.4   Neuron simulation

Neuron circuit explained in section 5.2, has been simulated with input resistance of 500 MΩ and input spike of 1 V. During integration, V_{fb} remains at 2.5 V and input spike of low voltage 2.5 V, high voltage 3.5 V, frequency of 100 Hz and duty cycle of 50 % has been applied at the spike input. Each incoming high spike generates a spike current of 2 nA. During integration time V_{ph} is low setting neuron in integration phase. In this phase neuron integrates spike current in 235 pF capacitor (C). Calculated integrated time

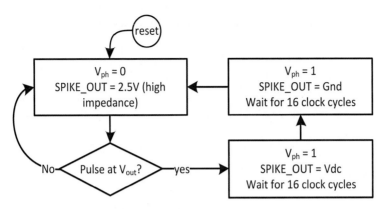

<figure>FIGURE 5.21: Spike Generator algorithm [63]</figure>

as per equation 5.28 is $470\ ms$. As shown in simulation (figure 5.3), when V_{ph}(orange) is low, *input spikes* (magenta) generates current and V_{op} (green) voltage goes low from 2.5 v till V_{th} (pink) (1 V). When V_{op} reaches 1 V , comparator triggers and spike generator circuit activates. Spike generator circuit goes high (5 V) for 16 clock cycles of $312.5\ \mu s$ clock period and then goes low (0 V) for next 16 clock cycle as shown in figure 5.22 and 5.3.

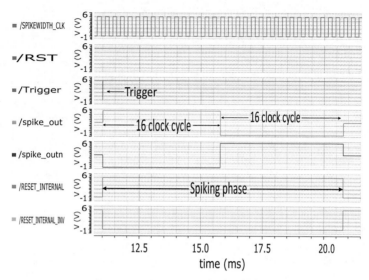

FIGURE 5.22: Simulation result of spiking circuit

5.5   Neuron measurement

The response of the integrate and fire neuron was tested by applying constant input current pulses and measuring the integration time needed until the circuit signals a fire event as shown in Fig. 5.23 for a specific set of input signals. The resting potential (V_{fb}) is $2.5\,\text{V}$ as explained in 5.3.3 , the threshold was set to $1.5\,\text{V}$ while a voltage pulse (low level = $2.5V$, high level = $3.5V$, frequency $100Hz$ and duty cycle = 50 %) was applied at *input spike* (fig. 5.2).

The integration time can be calculated as

$$t = \frac{C \cdot (V_{th} - V_{fb})}{I_{in,avg}} \tag{5.28}$$

where C is the feedback cap, V_{th} is threshold voltage, V_{fb} is resting potential and $I_{in,avg}$ is input current.

94

Theoretical integration time (**t**) calculated for this set of input is 469 ms and measurement shows the integration time of 470 ms as shown in figure 5.23. Measurement results show a good agreement with theoretical calculation.

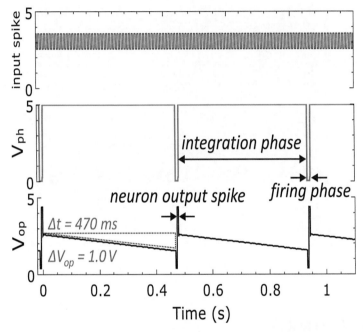

FIGURE 5.23: Neuron spiking measurement for constant average input current [63]

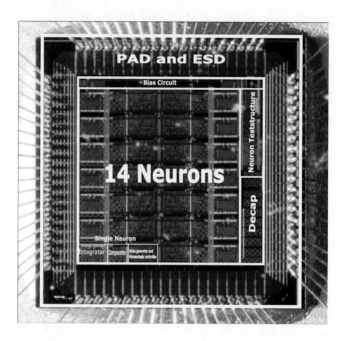

FIGURE 5.24: Neuron ASIC fabricated in AMS 350 nm process. Each ASIC accommodates 14 spiking neuron circuit. Neuron Testbench in the figure has all pins accessible externally for testing purpose. Die has been bond wired in QFN80 package [63]

.

5.6 Conclusion

An integrate and fire neuron circuit with spiking feature has been implemented and tested in AMS 350 nm process. Similar to biological neuron, CMOS neuron shows integration and spiking phase. During integration phase neuron integrates all incoming currents and it spikes during spiking phase when neuron reaches threshold voltage. Neuron spike appears at input and output simultaneously. The integration phase duration depends on the input current and feedback capacitor. The spiking pulse duration is externally adjustable and can be controlled by frequency of input clock. Adjustable duration of pulse width provides the freedom to control the speed of learning in memristor based network.

Chapter 6

Neuron signaling and memristor integration with neuron ASIC

A biological brain has a remarkable property of adaptation. A brain can acquire, coordinate and process input signals through five sensory organs and take intelligent decisions which are even impossible for modern days digital computers and also these computers do not work efficiently like our brain does. Our brain can process the information in milliseconds and store the data for years. Many kinds of research are going on to implement efficient neural network on the hardware. A memristive neural network is a mimicking of biological neural system present in the human brain. Either it is an analog or digital or mixed signal circuit, it is important to understand the environment for the system to be designed so that its architecture and specifications can be decided. In this chapter we first explain the neurophysiology, then the integration of memristor with electronic neuron to mimic the fundamental unit of a brain which can be repeated and interconnected in billions to develop the biological brain functionalities. The fundamental component of a neural network are neurons and synapse (interconnect between neurons). In our human brain we have 100 billion neurons and 1000 trillion synapses. Each neuron is connected with 10000 neurons through these synapses.

6.1 Neuron signaling

Neurons transmit information by generating electrical spikes. Neurons are not a good conductor of electricity but they have evolved chemical ions based elaborate mechanism for signal transmission over long distances. Fundamental components of neuron are cell body, dendrites, axon and synapses as shown in figure 6.1. A cell membrane is a semipermeable membrane surrounding the cytoplasm of a cell. This isolates inside of the cell from outside. At the default state, neurons membrane potential remain $-70\ mV$ which means membrane potential inside neuron is negative as compared to outside. During an action potential, the membrane potential temporarily goes positive up to $+40\ mV$. These action potential propagate from one neuron to other via axon. These are the fundamental signals which carry the information. The electrical signals produced by a neuron is due to the stimuli which it receives from multiple neurons through synaptic cleft or directly by the sensory organs like light, sound or touch. The amplitude of these electrical signal (action potential or spikes) remains the same and is independent of the input stimuli strength. The information is encoded in the frequency of spike, not the amplitude.

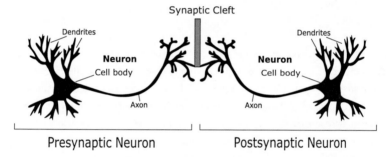

FIGURE 6.1: Neuron and synapses

Neurons communicate at synaptic cleft with the movement of calcium (Ca^{2+}), sodium ion (Na^+) and neurotransmitter as shown in figure 6.2. Neurons have a higher concentration of potassium ions inside and a higher percentage of sodium ions outside. On the membrane surface, neurons have a dedicated leak channel for potassium and sodium ions. When an action potential arrives at the presynaptic neuron, it stimulates the opening of the voltage-gated channel for Ca^{2+}. Ca^{2+} diffuses into the cytoplasm of a presynaptic membrane. Ca^{2+} cause vesicles containing acetylcholine neurotransmitter to move towards the presynaptic membrane. The vesicle fuses with the membrane, and thus neurotransmitter is released and then it diffuses across the synaptic cleft, and it temporarily binds to receptor proteins on the postsynaptic membrane causing the chemically gated Na^+ channel to open. Na^+ diffuses through the postsynaptic membrane and increases potential inside postsynaptic neuron which may cause action potential or support in an action potential. When inside membrane potential reaches $-55\ mV$, it opens the voltage-gated sodium channel which allows more sodium ions to flow inside which increases membrane potential up to $+30$ mV. When the membrane potential reaches $+30$ mV, then the potassium channel opens and allows more potassium ions to move outwards which reduces the membrane potential and is called repolarization as shown in figure 6.2. Since potassium channel takes some time to close, it allows the potassium ions to flow out resulting in hyperpolarization. Later the membrane potential settles again at -70 mV due to leak channels.

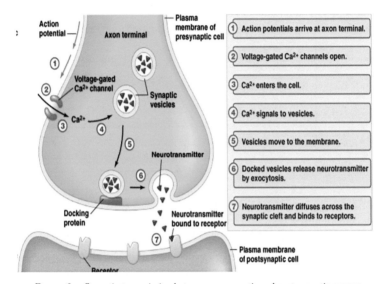

FIGURE 6.2: Synaptic transmission between pre-synaptic and post-synaptic neuron

6.2 Memristor and neuron ASIC integration

As explained in section 6.1, synapse and neurons are the fundamental units of the bigger neural network. In this section the details of double-barrier memristor device are explained that is specifically developed for integration with the neuron circuit.

6.2.1 Double barrier memristor

The double barrier memristive device developed at Kiel University consists of an aluminum oxide tunnel barrier and a Schottky barrier at the niobium oxide / gold interface as

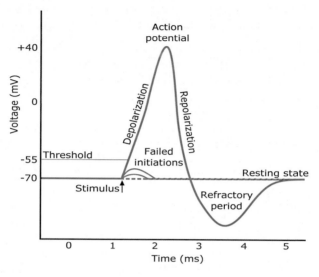

FIGURE 6.3: Action Potential [112]

shown in figure 6.4. Al_2O_3 and Nb_xO_y has thickness of $1.3\ nm$ and $2.5\ nm$ respectively. There are two different physical mechanisms describing the memristive characteristics of double barrier devices. Figure 6.4 (a) shows the first mechanism where Nb_xO_y layer acts as trapping layer for electrons. The localized electronic states within Nb_xO_y are emptied or filled depending on the applied voltage polarity and duration. So, the amount of charge displacement depends upon the history of the applied voltage and charged trap represents a high resistance state whereas discharged trap represents a low resistance state.

There is another physical mechanism on contrary to charge injection model shown in figure 6.4 (a). In this mechanism, Nb_xO_y represents a solid-state electrolyte and Al_2O_3 acts like a tunnel barrier. When a bias voltage is applied, oxygen ions (within the Nb_xO_y) drifts toward Al_2O_3 which affects interfacial parameters at Al_2O_3 / Nb_xO_y and Nb_xO_y/ Au (Schottky) interface. These ions movements alter device resistance leading to the memristive hysteresis I-V curve shown in figure 6.5. Detailed description of the fabrication, electrical characterization and switching mechanism of DBMD can be found in [51], [64].

"The double barrier memristive device was developed for neuromorphic applications, and has several key benefits, which facilitate the device integration. No current limitation

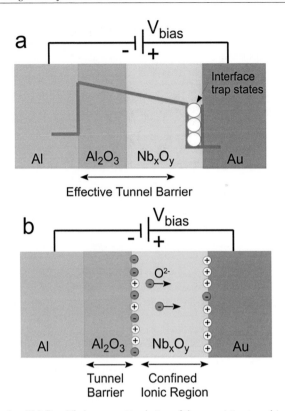

FIGURE 6.4: "(a) Simplified cross-sectional view of the memristive tunnel junctions. Here, trap states within the NbxOy are assumed. The filling and emptying of traps by injected electrons varies the amount of charge in the NbxOy layer and therefore the resistance. (b) An alternative model to (a). Under forward bias voltages Vbias oxygen ions (orange circles) can move inside the NbxOy layer, where their diffusion region is confined by the Al2O3 layer and the NbxOy/Au interface. Both, the model in (a) as well as the model in (b) describe the memristive I–V characteristics." [51].

is necessary for the set- and reset-process and the resistance changes gradually. This allows to precisely control the resistance of the device, with exponential dependencies on the pulse amplitude and duration for the set- and reset process [51]. Furthermore, the devices were integrated into passive crossbar arrays with 256 devices (Fig. 6.6). and no selector devices are necessary because of the high asymmetry of the I-V-curve. The absence of a distinct electroforming step further simplifies the usage and keeps the device-to-device variation to a minimum [51]. " [63]

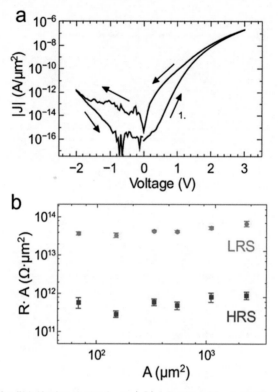

FIGURE 6.5: "(a) Absolute current density | J | as function of the applied bias voltage. (b) The area-resistance product vs. junction-area curve of the double barrier device measured at 0.5 V indicates a homogeneous area dependent charge transport. The error bars are obtained from 5 cells of each area. Junction areas were confirmed with optical microscopy. "[51]

6.2.2 Memristor and neuron integration circuit

A neuromorphic circuit includes the dense structure of memristor and I & F neuron. Each unit has property to receive incoming currents through memristor (like a synapse in the biological neuron), integrates it in neuron and fire when neuron reach the threshold. The memristor die used has been shown in figure 6.6. This die contains 256 devices in 16 X 16 array. The die shows a wide spread of off resistance as shown in figure 6.7. Since the die shows huge variation in off resistance, the memristor of row 4 and column 15 has been used as its resistance is 2 GΩ and it fits for integration with the designed neuron.

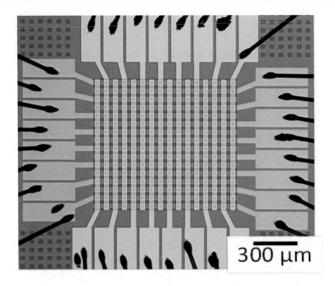

FIGURE 6.6: Image of the wirebonded crossbar array with 256 memristive devices [63]

The memristor (R_4C_{15}) has been connected with CMOS neuron ASIC as shown in fig. 6.8. Negative terminal of memristor is connected with negative input of integrator and input spike is given at positive terminal of memristor. Input spikes of $V_{low} = 2.5V$, $V_{high} = 3V$, freq = 40 Hz and duty cycle = 50 % is applied (red spike in figure 6.8). Since the positive terminal of integrator is at $V_{fb} = 2.5V$ which is default state of operation, due to internal regulation of op-amp V_{In} remains at 2.5 V. Each spike makes a voltage difference of $V_{diff} = 0.5\ V$ across memristor and current of (equation 6.1).

$$I_{memristor} = \frac{0.5V}{R_{memristor}} \tag{6.1}$$

This allows an input current for the integrator during the pulse time of 12.5 ms. Each spike integrates the incoming currents in integrator and fires when output of integrator (V_{op}) reaches the comparator threshold (V_{th}) of 1.5 V. At this state, comparator triggers digital spike circuit which generates negative (0 V) for 5 ms and positive pulse (5 V) for 5 ms and then settles at default voltage level of 2.5 V. The spike appears at V_{fb} node of the integrator and negative terminal of memristor (in).

If output spike appears during the time when input spike is present, it develops a higher voltage drop (write voltage > 3 V) across memristor which changes the resistance and

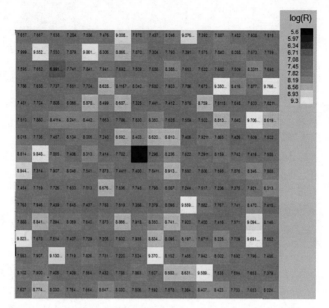

FIGURE 6.7: Off resistance variation in 16 X 16 die

memristor gets more conducting in steps due to ion movement as explained in [51] and section 6.2.1.

During the time no voltage is applied (off time of pulse when the voltage level is $2.5\ V$), memristor retains the current conductance state. Each spike makes memristor more conductive allowing more average input current for integration. In this state neuron fires at a higher rate which happens due to a permanent change in the resistance of memristor. The device may have retention issues to retain the non-volatile memory for a longer time which will be explained in 6.3.

6.2.3   Memristive system measurement

The input spiking explained in section 5.2 has been used for integration (shown in figure 5.2). The red input spikes (yellow in figure 5.2) with specification explained in section 6.2.2 is applied with a pulse train which integrates incoming current (equation 6.1) and output voltage of integrator (V_{op}) reduces voltage until it reaches threshold of 1.5 V as

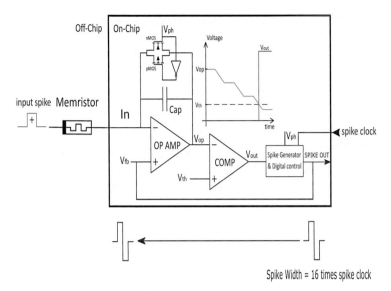

FIGURE 6.8: Neuron circuit with memristor [63]

shown in figure 6.9. At this stage, a digital circuit is triggered and it generates a spike which appears at the negative terminal of the memristor. It takes $37s$ in the beginning which shows the highest resistance state of the memristor.

This resistance reduces slowly and firing time reduces till $3.5s$. The output has been measured using an oscilloscope (PicoScope 3000 series). It can be seen that memristor shows 10 times change of resistance after 20 potentiation spikes.

This resistance change shows a learning behavior of memristor on hardware circuit.

6.3 Memristor retention issue

Memristor shows learning behavior which makes it a promising candidate for neuromorphic circuits but with issues like reproducibility, retention etc. It remains a technological challenge to fabricate memristor with good retention which qualifies memristor for non-volatile memories. A retention test has been conducted on memristor to study the relaxation behavior of resistance state.

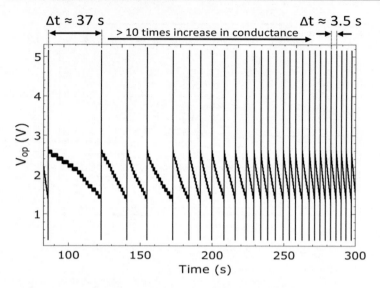

FIGURE 6.9: Neuron circuit output measurement [63]

The system has been potentiated for 750 pulses which shows 100 times change in conductance as shown in figure 6.10. At starting, the memristor is at highest resistance state and the firing period is very long of 110 seconds. This firing period decreases as memristor gets more conductive and eventually reaches to firing period of 1 second after input spike of 2000 seconds. To investigate retention of the memristive device, applied input spike voltage was turned off for 300 seconds and spiking period was measured. Spiking period has changed from 1 second to 3 seconds as shown in figure 6.10 which shows 3 times change in conductance in 300 seconds. Input pulses were applied again for 200 second which shows a change in firing rate again till maximum conductance and firing period of 1 second. The process was again repeated and spike input was turned off for 600 second and the change in firing rate was the same as for 300 seconds. A slight increase in firing period is observed in both cases but it quickly reaches previous firing rate when input pulses applied. Since the increase in firing rate is similar in both cases, it can be assumed that relaxation occurs in the first few seconds and then resistance increase slows down.

This relaxation limits the application of memristor device to be used in memory applications, but it can be used in the neuromorphic application where this much relaxation may be less dramatic. Considering that the requirements for future neuromorphic applications are not yet clear, the moderate increase in resistance may not be a disadvantage but a useful behavior, since it allows to mimic transient neural dynamics such as spike frequency adaptation [113], [114], [115].

In 1982, Oja [116] proposed a learning rule in which coupling growth shows asymptotic stability due to forgetting rate which becomes stronger with an increase in synaptic efficacy. This overcomes the problem of unbound growth in synaptic coupling strength in several neuron models. The asymptotic decrease of the firing period (i.e. the device resistance) shows similar behavior as mentioned in Oja's rule and this result can be a hardware implementation of Oja's rule.

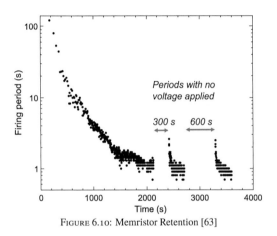

FIGURE 6.10: Memristor Retention [63]

6.4 Conclusion

DBMD shows pinched hysteresis characteristics operating at ultra low current which makes it a suitable choice for low power high density memristive system. DBMD devices shows wide variation over wafer and this remains as a challenge for this technology to develop devices with less deviation across wafer. This device has been integrated with CMOS neuron circuit and measurement results shows 100 times change in memristor resistance. However, it shows fast relaxations at low resistance state which limits this device application in memory but suits for neuromorphic applications where this relaxation may not be a disadvantage. Integrated DBMD and CMOS neuron behaves as fundamental element for neuromorphic computing which paves the way towards large neuromorphic system.

Chapter 7

Conclusion and outlook

Memristor-based neuromorphic circuits are considered to be more efficient than present-day digital computer because of its distributed computing and memory storage in networks. Many researchers have used different materials and architectures to develop memristors with a different current-voltage relationship. However, there is still ambiguity about the best memristive device for neuromorphic applications. Emulator ASIC designed in this work to address this issue has a programmable feature which allows implementing any memristive function on hardware to observe memristive behavior. Key components of biological learning are LTP, LTD and synaptic plasticity which has been tested on memristor emulator and shows brain like learning behavior in hardware. Moreover, this emulator is compact compared to other emulators developed with commercial components and it also provides a platform to develop neuromorphic systems on the hardware level and modify the network via external programming.

Memristive based networks are capable of processing a high volume of data which remains a challenge for modern-day computers. In this work, memristor-based pattern recognition has been simulated in LTspice with the model of real devices and the system is found to be capable of performing unsupervised learning. This network has the possibility to extend to any bigger array and can be used for high-resolution images processing. The memristor-based approach does not suffer from communication bandwidth and memory wall issues which is a serious challenge for modern day digital computers.

Development of memristor-based neuromorphic computers includes the development of memristor, CMOS circuits, and its integration to develop NPU. In this work, an NPU has been developed with DBMD and neuron ASIC. DBMD developed at Kiel university shows continuous resistance change and has better retention. However, memristor suffers from reproducibility and device variation which will improve as memristor fabrication technology mature with time. CMOS neuron circuit developed in AMS350 nm process shows the fundamental behavior of a biological neuron. CMOS neurons are capable of integrating currents through the memristor, compare its voltage level with neuron threshold and generates a spike similar to a biological action potential.

Integration of memristor with neuron ASIC develops the NPU which resembles like synapse-neuron in the brain. This unit can be repeated in any neuromorphic application to develop cognitive task on hardware which has been in simulation till now like in [62], [45], [41], [117] etc. NPU developed in this work shows LTP in measurement which shows learning capability of memristor on hardware. The change in memristor is continuous which provides high resolution of memory in single device rather than binary switching memristors [15], [16], [17]. Hence, DBMD and NPU developed in this work shows to be a good candidate for neuromorphic applications.

Future work

This section details improvement of neuromorphic circuits and implementation of memristor and neuron unit in a bigger network for implementing brain like functionalities on hardware. DBMD fabrication technology has to be improved for better retention, reproducibility, and variability. This will allow this device to be useful not only in short-term memory application but long-term memory applications as well. Once, the technology is mature, NPU can be integrated in high density to develop memristor-CMOS based neuromorphic computers as proposed in figure 7.1. The proposed neuromorphic computer has NPU as the fundamental unit which is repeated in a 3D array with input layer, output layer, and network and programming layer. Each neuron output and memristor input is connected with output layer as well as network and programming layer which provides each memristor to be programmed individually. Input layer provides input signal which may be an encoded audio, video or any other signal to be processed. This architecture has highly parallel architecture and can be extended to any limit according to application.

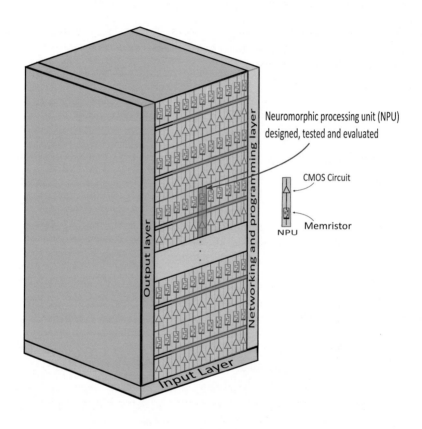

FIGURE 7.1: Neuromorphic processor model consisting of NPU in 3D array, input layer, output layer, networking and programming layer. Architecture can be extended in massively parallel array for high processing applications.

Appendix A

Description of emulator ASIC pins used for testing purpose

TABLE A.1: Memristor Emulator ASIC pin assignments for QFN64

Pin	Name	Description
1	vdd!	Power supply 3.3V
2	VDDA	Power supply 3.3V
3	GNDA	Virtual Ground 1.65V
4	gnd!	Ground Supply
5	VSSA	Ground Supply
6	PAD_Emul_1_eoc	End of conversion output of voltage measurement for 400K to 204.8M emulator.
7	PAD_Emul_1_serial_out	Serial output data of measured voltage across emulator
8	PAD_Neuron_1_out	Comparator Output of first neuron
14	PAD_Neuron_1_vth	Comparator threshold of first neuron
15	PAD_Neuron_1_ph	Voltage input for switching between the integration and spiking phase of 1st neuron
16	PAD_Neuron_1_fb	Feedback for 1st neuron
17	PAD_Neuron_1_In	Spike input for 1st neuron
18	PAD_Emul_4_eoc	End of conversion output of voltage measurement for 10K to 5P12M emulator
19	PAD_Emul_4_serial_out	Serial output data of measured voltage across emulator(10k to 5.12M)

Continued on next page

111

Table A.1 – *Continued from previous page*

Pin	Name	Description
22	PAD_Emul_1_Clk_ADC	ADC clock for 400k to 204.8M emulator
23	PAD_Emul_1_Clk_P_S	Parallel to serial clock for ADC output of emulator(400k to 204.8M)
24	PAD_Emul_SC_Clk_P_S	Parallel to serial clock for ADC output of emulator(switch cap).
25	PAD_Emul_SC_Clk_ADC	ADC clock for switch cap emulator.
26	PAD_Emul_4_Clk_P_S	Parallel to serial clock for ADC output of emulator(10k to 5.12M)
30	PAD_Emul_SC_1_clk	Input Clock for 1st switch cap emulator.
31	PAD_Emul_4_1_global_reset	Global reset for serial to parallel converter digital block
32	PAD_Emul_SC_Start_ADC	Start ADC signal for switch capacitor emulator.
33	PAD_Emul_SC_analog_mux_S1	LSB for switch cap analog mux.
34	PAD_Emul_SC_analog_mux_S0	MSB for switch cap analog mux.
35	PAD_Emul_4_new_sample	New sample signal for the input of new serial data in 10k to 5.12M emulator
36	PAD_Emul_4_data_in	Serial Input data for 10k to 5.12M emulator
37	PAD_Emul_4_clk_in	Input clock for the serial data(10k to 5.12M)
38	PAD_Emul_4_analog_mux_S1	LSB for 10k to 5.12M analog mux
39	PAD_Emul_4_analog_mux_S0	MSB for 10k to 5.12M analog mux
40	PAD_Emul_4_Start_ADC	Start ADC signal for 10k to 5.12M emulator
41	PAD_Emul_4_DIG_IN_CS1	MSB for 10k to 5.12M digital mux
42	PAD_Emul_4_DIG_IN_CS0	LSB for 10k to 5.12M digital mux
47	PAD_Emul_SC_1_input_B	Analog input for the B terminal of 1st SC emulator
48	PAD_Emul_SC_1_input_A	Analog input for the A terminal of 1st SC emulator
49	PAD_Emul_4_4_input_B	Analog input for the B terminal of 4th 10k to 5.12M emulator
50	PAD_Emul_4_4_input_A	Analog input for the A terminal of 4th 10k to 5.12M emulator

Continued on next page

Table A.1 – *Continued from previous page*

Pin	Name	Description
51	PAD_Emul_4_3_input_B	Analog input for the B terminal of 3rd 10k to 5.12M emulator
52	PAD_Emul_4_3_input_A	Analog input for the A terminal of 3rd 10k to 5.12M emulator
53	PAD_Emul_4_2_input_B	Analog input for the B terminal of 2nd 10k to 5.12M emulator
54	PAD_Emul_4_2_input_A	Analog input for the A terminal of 2nd 10k to 5.12M emulator
55	PAD_Emul_4_1_input_B	Analog input for the B terminal of 1st 10k to 5.12M emulator
56	PAD_Emul_4_1_input_A	Analog input for the A terminal of 1st 10k to 5.12M emulator
57	PAD_Emul_1_input_B	Analog input for the B terminal 400k to 204.8M emulator
58	PAD_Emul_1_input_A	Analog input for the A terminal 400k to 204.8M emulator
59	PAD_Emul_ALL_PD_Analog	Power down signal for all analog block
60	PAD_Emul_1_New_sample_serial_in	New sample signal for the input of new serial data in 400k to 204.8M emulator
61	PAD_Emul_1_Serial_data_in	Serial Input data for 400k to 204.8M emulator
62	PAD_Emul_1_Serial_data_in_clk	Input clock for the serial data input(400k to 204.8M)
63	PAD_Emul_1_Start_ADC	Start ADC signal for 400k to 204.8M emulator

113

Appendix B

Description of neuron ASIC pins used for testing purpose

TABLE B.1: Neuron ASIC pin assignments for QFN80

Pin	Name	Description
52	PAD_TB_NEURON_ID_OUT	Neuron ID output of test-bench
53	PAD_TESTBENCH_NEURON_ID_IN	Neuron ID input of test-bench
54	PAD_TESTBENCH_ENABLE_ID_WRITE	Write enable signal for Id writing of each neuron
55	PAD_TESTBENCH_SPIKE_WIDTH_CLK	Clock for spike width of testbench neuron
56	PAD_TESTBENCH_NEURON_ID_CLK_LOWF	Low frequency clock for Homeostasis
58	PAD_TESTBENCH_NEURON_ID_CLK_HIGHF	High frequency clock for Homeostasis
61	VDD5V_ANALOG	5 V supply for analog blocks
62	GND5V_ANALOG	Ground supply for analog blocks
63	PAD_TESTBENCH_RST	Reset signal for neuron testbench

Continued on next page

Table B.1 – *Continued from previous page*

Pin	Name	Description
64	PAD_TB_SPIKE_OUT	Output of neuron test-bench
65	PAD_TESTBENCH_NEURON_SPIKE_IN	Input for neuron testbench
66	PAD_TB_INT_OUT	Testbench neuron integrator output
67	PAD_TB_RESET_ALL	Reset for neuron testbench
68	V_GND	Virtual ground (2.5 V) signal for all neuron
69	PAD_TB_NEURON_VTH	Threshold signal for neuron testbench
71	gnd!	Ground supply of digital
72	vdd!	5 V supply for digital block
73	PAD_NEURON_ID_OUT	Output for neuron id
79	VDD_3P3	3.3 V power supply
80	GND_3P3	3.3 V ground supply

Bibliography

[1] Maurice V Wilkes. The memory wall and the cmos end-point. *ACM SIGARCH Computer Architecture News*, 23(4):4–6, 1995.

[2] Giacomo Indiveri, Bernabé Linares-Barranco, Robert Legenstein, George Deligeorgis, and Themistoklis Prodromakis. Integration of nanoscale memristor synapses in neuromorphic computing architectures. *Nanotechnology*, 24(38): 384010, 2013.

[3] Teresa Serrano-Gotarredona, Timothée Masquelier, Themistoklis Prodromakis, Giacomo Indiveri, and Bernabe Linares-Barranco. Stdp and stdp variations with memristors for spiking neuromorphic learning systems. *Frontiers in neuroscience*, 7:2, 2013.

[4] Leon O Chua. Memristor-the missing circuit element. *Circuit Theory, IEEE Transactions on*, 18(5):507–519, 1971.

[5] Dmitri B Strukov, Gregory S Snider, Duncan R Stewart, and R Stanley Williams. The missing memristor found. *Nature*, 453(7191):80–83, 2008.

[6] Patrick WC Ho, Haider AF Almurib, and T Nandha Kumar. One-bit non-volatile memory cell using memristor and transmission gates. In *Electronic Design (ICED), 2014 2nd International Conference on*, pages 244–248. IEEE, 2014.

[7] Myoung-Jae Lee, Chang Bum Lee, Dongsoo Lee, Seung Ryul Lee, Man Chang, Ji Hyun Hur, Young-Bae Kim, Chang-Jung Kim, David H Seo, Sunae Seo, et al. A fast, high-endurance and scalable non-volatile memory device made from asymmetric ta 2 o 5- x/tao 2- x bilayer structures. *Nature materials*, 10(8):625, 2011.

[8] Jinjoo Park, Seunghyup Lee, Junghan Lee, and Kijung Yong. A light incident angle switchable zno nanorod memristor: reversible switching behavior between two non-volatile memory devices. *Advanced Materials*, 25(44):6423–6429, 2013.

[9] Rainer Waser, Regina Dittmann, Georgi Staikov, and Kristof Szot. Redox-based resistive switching memories–nanoionic mechanisms, prospects, and challenges. *Advanced materials*, 21(25-26):2632–2663, 2009.

[10] Yuriy V Pershin and Massimiliano Di Ventra. Practical approach to programmable analog circuits with memristors. *Circuits and Systems I: Regular Papers, IEEE Transactions on*, 57(8):1857–1864, 2010.

[11] Sangho Shin, Kyungmin Kim, and Sung-Mo Kang. Memristor applications for programmable analog ics. *IEEE Transactions on Nanotechnology*, 10(2):266–274, 2011.

[12] Tian Xiao-Bo and Xu Hui. The design and simulation of a titanium oxide memristor-based programmable analog filter in a simulation program with integrated circuit emphasis. *Chinese Physics B*, 22(8):088501, 2013.

[13] Carlos Zamarreño-Ramos, L Camuñas-Mesa, Jose A Pérez-Carrasco, Timothée Masquelier, Teresa Serrano-Gotarredona, and Bernabé Linares-Barranco. On spike-timing-dependent-plasticity, memristive devices, and building a self-learning visual cortex. *Frontiers in Neuroscience*, 5(00026), 2011.

[14] Finn Zahari, Mirko Hansen, Thomas Mussenbrock, Martin Ziegler, and Hermann Kohlstedt. Pattern recognition with tio x-based memristive devices. *AIMS Materials Science*, 2:2372–0484, 2015.

[15] H Pagnia and N Sotnik. Bistable switching in electroformed metal–insulator–metal devices. *physica status solidi (a)*, 108(1):11–65, 1988.

[16] S d Q Liu, NJ Wu, and Alex Ignatiev. Electric-pulse-induced reversible resistance change effect in magnetoresistive films. *Applied Physics Letters*, 76(19):2749–2751, 2000.

[17] Y Watanabe, JGe Bednorz, A Bietsch, Ch Gerber, D Widmer, A Beck, and SJ Wind. Current-driven insulator–conductor transition and nonvolatile memory in chromium-doped srtio 3 single crystals. *Applied Physics Letters*, 78(23): 3738–3740, 2001.

[18] Haruki Toda. Phase change memory device, February 26 2008. US Patent 7,335,906.

[19] Young-Nam Hwang and Young-Tae Kim. Phase-change memory devices, June 27 2006. US Patent 7,067,837.

[20] Osama Khouri and Claudio Resta. Phase change memory device, May 23 2006. US Patent 7,050,328.

[21] Ping Zhou, Bo Zhao, Jun Yang, and Youtao Zhang. Energy reduction for stt-ram using early write termination. In *Proceedings of the 2009 International Conference on Computer-Aided Design*, pages 264–268. ACM, 2009.

[22] Clinton W Smullen, Vidyabhushan Mohan, Anurag Nigam, Sudhanva Gurumurthi, and Mircea R Stan. Relaxing non-volatility for fast and energy-efficient stt-ram caches. In *High Performance Computer Architecture (HPCA), 2011 IEEE 17th International Symposium on*, pages 50–61. IEEE, 2011.

[23] KL Wang, JG Alzate, and P Khalili Amiri. Low-power non-volatile spintronic memory: Stt-ram and beyond. *Journal of Physics D: Applied Physics*, 46(7): 074003, 2013.

[24] BN Engel, J Akerman, B Butcher, RW Dave, M DeHerrera, M Durlam, G Grynkewich, J Janesky, SV Pietambaram, ND Rizzo, et al. A 4-mb toggle mram based on a novel bit and switching method. *IEEE Transactions on Magnetics*, 41(1):132–136, 2005.

[25] Said Tehrani, JM Slaughter, E Chen, M Durlam, J Shi, and M DeHerren. Progress and outlook for mram technology. *IEEE Transactions on Magnetics*, 35(5):2814–2819, 1999.

[26] Shoichiro Kawashima, Toru Endo, Akira Yamamoto, Ken'ichi Nakabayashi, Mitsuhara Nakazawa, Keizo Morita, and Masaki Aoki. Bitline gnd sensing technique for low-voltage operation feram. *IEEE Journal of Solid-State Circuits*, 37(5):592–598, 2002.

[27] T Mikolajick, C Dehm, W Hartner, I Kasko, MJ Kastner, N Nagel, M Moert, and C Mazure. Feram technology for high density applications. *Microelectronics Reliability*, 41(7):947–950, 2001.

[28] Jagan Singh Meena, Simon Min Sze, Umesh Chand, and Tseung-Yuen Tseng. Overview of emerging nonvolatile memory technologies. *Nanoscale research letters*, 9(1):526, 2014.

[29] Mohamed T Ghoneim and Muhammad M Hussain. Neuron-inspired flexible memristive device on silicon (100). *arXiv preprint arXiv:1706.05645*, 2017.

[30] Can Li, Lili Han, Hao Jiang, Moon-Hyung Jang, Peng Lin, Qing Wu, Mark Barnell, J Joshua Yang, Huolin L Xin, and Qiangfei Xia. Three-dimensional crossbar arrays of self-rectifying si/sio 2/si memristors. *Nature Communications*, 8:15666, 2017.

[31] H-S Philip Wong, Heng-Yuan Lee, Shimeng Yu, Yu-Sheng Chen, Yi Wu, Pang-Shiu Chen, Byoungil Lee, Frederick T Chen, and Ming-Jinn Tsai. Metal–oxide rram. *Proceedings of the IEEE*, 100(6):1951–1970, 2012.

[32] Gina C Adam, Brian D Hoskins, Mirko Prezioso, Farnood Merrikh-Bayat, Bhaswar Chakrabarti, and Dmitri B Strukov. 3-d memristor crossbars for analog and neuromorphic computing applications. *IEEE Transactions on Electron Devices*, 64(1):312–318, 2017.

[33] Pengxiao Sun, Nianduan Lu, Ling Li, Yingtao Li, Hong Wang, Hangbing Lv, Qi Liu, Shibing Long, Su Liu, and Ming Liu. Thermal crosstalk in 3-dimensional rram crossbar array. *Scientific reports*, 5:13504, 2015.

[34] Yuan Heng Tseng, Chia-En Huang, C-H Kuo, Y-D Chih, and Chrong Jung Lin. High density and ultra small cell size of contact reram (cr-ram) in 90nm cmos logic technology and circuits. In *Electron Devices Meeting (IEDM), 2009 IEEE International*, pages 1–4. IEEE, 2009.

[35] Chia-Fu Lee, Hon-Jarn Lin, Chiu-Wang Lien, Yu-Der Chih, and Jonathan Chang. A 1.4 mb 40-nm embedded reram macro with 0.07 um 2 bit cell, 2.7 ma/100mhz low-power read and hybrid write verify for high endurance application. In *Solid-State Circuits Conference (A-SSCC), 2017 IEEE Asian*, pages 9–12. IEEE, 2017.

[36] Carver Mead. *Analog VLSI and Neural Systems*. Addison-Wesley Longman Publishing Co., Inc., Boston, MA, USA, 1989. ISBN 0-201-05992-4.

[37] Rodney Douglas, Misha Mahowald, and Carver Mead. Neuromorphic analogue vlsi. *Annual review of neuroscience*, 18(1):255–281, 1995.

[38] Federico Faggin and Carver Mead. Vlsi implementation of neural networks. 1990.

[39] Carver Mead. Neuromorphic electronic systems. *Proceedings of the IEEE*, 78 (10):1629–1636, 1990.

[40] Ronald Tetzlaff. *Memristors and memristive systems*. Springer, 2013.

[41] Damien Querlioz, Olivier Bichler, and Christian Gamrat. Simulation of a memristor-based spiking neural network immune to device variations. *Neural Networks (IJCNN), The 2011 International Joint Conference on*, pages 1775–1781, 2011.

[42] José Antonio Pérez-Carrasco, Carlos Zamarreno-Ramos, Teresa Serrano-Gotarredona, and Bernabé Linares-Barranco. On neuromorphic spiking architectures for asynchronous stdp memristive systems. In *Circuits and Systems (IS-CAS), Proceedings of 2010 IEEE International Symposium on*, pages 1659–1662. IEEE, 2010.

[43] Chris Yakopcic and Tarek M Taha. Energy efficient perceptron pattern recognition using segmented memristor crossbar arrays. In *Neural Networks (IJCNN), The 2013 International Joint Conference on*, pages 1–8. IEEE, 2013.

[44] Yuanfan Yang, Jimson Mathew, and Dhiraj K Pradhan. Matching in memristor based auto-associative memory with application to pattern recognition. In *Signal Processing (ICSP), 2014 12th International Conference on*, pages 1463–1468. IEEE, 2014.

[45] Mirko Hansen, Finn Zahari, Martin Ziegler, and Hermann Kohlstedt. Double-barrier memristive devices for unsupervised learning and pattern recognition. *Frontiers in neuroscience*, 11:91, 2017.

[46] Patrick Sheridan, Wen Ma, and Wei Lu. Pattern recognition with memristor networks. In *Circuits and Systems (ISCAS), 2014 IEEE International Symposium on*, pages 1078–1081. IEEE, 2014.

[47] Yang Zhang, Yi Li, Xiaoping Wang, and Eby G Friedman. Synaptic characteristics of ag/aginsbte/ta-based memristor for pattern recognition applications. *IEEE Transactions on Electron Devices*, 64(4):1806–1811, 2017.

[48] Erika Covi, Stefano Brivio, Alexantrou Serb, Themistoklis Prodromakis, M Fanciulli, and S Spiga. Hfo2-based memristors for neuromorphic applications. In *Circuits and Systems (ISCAS), 2016 IEEE International Symposium on*, pages 393–396. IEEE, 2016.

[49] Son Ngoc Truong, Khoa Van Pham, Wonsun Yang, Kyeong-Sik Min, Yawar Abbas, and Chi Jung Kang. Live demonstration: Memristor synaptic array with fpga-implemented neurons for neuromorphic pattern recognition. In *Circuits and Systems (APCCAS), 2016 IEEE Asia Pacific Conference on*, pages 742–743. IEEE, 2016.

[50] Anik Chowdhury, Mrinmoy Sarkar, Aqeeb Iqbal Arka, and ABM Harun-ur Rashid. Associative memory algorithm for visual pattern recognition with memristor array and cmos neuron. In *Electrical and Computer Engineering (ICECE), 2016 9th International Conference on*, pages 42–45. IEEE, 2016.

[51] Mirko Hansen, Martin Ziegler, Lucas Kolberg, Rohit Soni, Sven Dirkmann, Thomas Mussenbrock, and Hermann Kohlstedt. A double barrier memristive device. *Scientific reports*, 5:13753, 2015.

[52] Antonio C Torrezan, John Paul Strachan, Gilberto Medeiros-Ribeiro, and R Stanley Williams. Sub-nanosecond switching of a tantalum oxide memristor. *Nanotechnology*, 22(48):485203, 2011.

[53] Mirko Prezioso, Farnood Merrikh-Bayat, BD Hoskins, GC Adam, Konstantin K Likharev, and Dmitri B Strukov. Training and operation of an integrated neuromorphic network based on metal-oxide memristors. *Nature*, 521(7550):61, 2015.

[54] Sung Hyun Jo, Kuk-Hwan Kim, Ting Chang, Siddharth Gaba, and Wei Lu. Si memristive devices applied to memory and neuromorphic circuits. In *Circuits and Systems (ISCAS), Proceedings of 2010 IEEE International Symposium on*, pages 13–16. IEEE, 2010.

[55] Qi Liu, Shibing Long, Hangbing Lv, Wei Wang, Jiebin Niu, Zongliang Huo, Junning Chen, and Ming Liu. Controllable growth of nanoscale conductive filaments in solid-electrolyte-based reram by using a metal nanocrystal covered bottom electrode. *ACS nano*, 4(10):6162–6168, 2010.

[56] Hyongsuk Kim, Maheshwar Pd Sah, Changju Yang, Seongik Cho, and Leon O Chua. Memristor emulator for memristor circuit applications. *IEEE Transactions on Circuits and Systems I: Regular Papers*, 59(10):2422–2431, 2012.

[57] Montree Kumngern. A floating memristor emulator circuit using operational transconductance amplifiers. In *Electron Devices and Solid-State Circuits (EDSSC), 2015 IEEE International Conference on*, pages 679–682. IEEE, 2015.

[58] Abdullah G Alharbi, Mohammed E Fouda, and Masud H Chowdhury. Memristor emulator based on practical current controlled model. In *Circuits and Systems (MWSCAS), 2015 IEEE 58th International Midwest Symposium on*, pages 1–4. IEEE, 2015.

[59] Antonio S Oblea, Achyut Timilsina, David Moore, and Kristy A Campbell. Silver chalcogenide based memristor devices. In *Neural Networks (IJCNN), The 2010 International Joint Conference on*, pages 1–3. IEEE, 2010.

[60] Chris Yakopcic, Andrew Sarangan, Jian Gao, Tarek M Taha, Guru Subramanyam, and Stanley Rogers. Tio 2 memristor devices. In *Aerospace and Electronics Conference (NAECON), Proceedings of the 2011 IEEE National*, pages 101–104. IEEE, 2011.

[61] Rajeev Ranjan, Pablo Mendoza Ponce, Anirudh Kankuppe, Bibin John, Lait Abu Saleh, Dietmar Schroeder, and Wolfgang H Krautschneider. Programmable memristor emulator asic for biologically inspired memristive learning. In *Telecommunications and Signal Processing (TSP), 2016 39th International Conference on*, pages 261–264. IEEE, 2016.

[62] Rajeev Ranjan, Pablo Mendoza Ponce, Wolf Lukas Hellweg, Alexandros Kyrmanidis, Lait Abu Saleh, Dietmar Schroeder, and Wolfgang H Krautschneider. Integrated circuit with memristor emulator array and neuron circuits for biologically inspired neuromorphic pattern recognition. *Journal of Circuits, Systems and Computers*, 26(11):1750183, 2017.

[63] R. Ranjan, M. Hansen, P. M. Ponce, L. A. Saleh, D. Schroeder, M. Ziegler, H. Kohlstedt, and W. H. Krautschneider. Integration of double barrier memristor die with neuron asic for neuromorphic hardware learning. In *2018 IEEE International Symposium on Circuits and Systems (ISCAS)*, pages 1–5, May 2018. doi: 10.1109/ISCAS.2018.8350996.

[64] Sven Dirkmann, Mirko Hansen, Martin Ziegler, Hermann Kohlstedt, and Thomas Mussenbrock. The role of ion transport phenomena in memristive double barrier devices. *Scientific reports*, 6:35686, 2016.

[65] Krzysztof Szot, Wolfgang Speier, Gustav Bihlmayer, and Rainer Waser. Switching the electrical resistance of individual dislocations in single-crystalline srtio 3. *Nature materials*, 5(4):312, 2006.

[66] Sieu D Ha and Shriram Ramanathan. Adaptive oxide electronics: A review. *Journal of applied physics*, 110(7):14, 2011.

[67] J Joshua Yang, Dmitri B Strukov, and Duncan R Stewart. Memristive devices for computing. *Nature nanotechnology*, 8(1):13, 2013.

[68] H Kohlstedt, K-H Gundlach, and S Kuriki. Electric forming and telegraph noise in tunnel junctions. *Journal of applied physics*, 73(5):2564–2568, 1993.

[69] A Baikalov, YQ Wang, B Shen, B Lorenz, S Tsui, YY Sun, YY Xue, and CWc Chu. Field-driven hysteretic and reversible resistive switch at the ag–pr 0.7 ca 0.3 mno 3 interface. *Applied Physics Letters*, 83(5):957–959, 2003.

[70] Evgeny Mikheev, Brian D Hoskins, Dmitri B Strukov, and Susanne Stemmer. Resistive switching and its suppression in pt/nb: Srtio 3 junctions. *Nature communications*, 5:3990, 2014.

[71] Seung Jae Baik and Koeng Su Lim. Bipolar resistance switching driven by tunnel barrier modulation in tio x/alo x bilayered structure. *Applied Physics Letters*, 97 (7):072109, 2010.

[72] Doo Seok Jeong, Byung-ki Cheong, and Hermann Kohlstedt. Pt/ti/al2o3/al tunnel junctions exhibiting electroforming-free bipolar resistive switching behavior. *Solid-State Electronics*, 63(1):1–4, 2011.

[73] *OP-LN-CMOS operational Amplifier*. Austria MicroSystems, 7 2015. Rev. A,29.07.2015.

[74] *ADC10A-CMOS 10-Bit ADC*. Austria MicroSystems, 7 2015. Rev. A,29.07.2015.

[75] S Singh, PWC Prasad, Abeer Alsadoon, A Beg, L Pham, and A Elchouemi. Survey on memrister models. In *Electronics, Information, and Communications (ICEIC), 2016 International Conference on*, pages 1–7. IEEE, 2016.

[76] Z. Biolek, D. Biolek, and V. Biolkova. Spice model of memristor with nonlinear dopant drift. 18(2):211, May 2009.

[77] Yogesh N Joglekar and Stephen J Wolf. The elusive memristor: properties of basic electrical circuits. *European Journal of Physics*, 2(4):1–24, 2009.

[78] Tim VP Bliss, Graham L Collingridge, et al. A synaptic model of memory: long-term potentiation in the hippocampus. *Nature*, 361(6407):31–39, 1993.

[79] Timothy J Teyler and P DiScenna. Long-term potentiation. *Annual review of neuroscience*, 10(1):131–161, 1987.

[80] Tim VP Bliss and Terje Lømo. Long-lasting potentiation of synaptic transmission in the dentate area of the anaesthetized rabbit following stimulation of the perforant path. *The Journal of physiology*, 232(2):331–356, 1973.

[81] Masao Ito. Long-term depression. *Annual review of neuroscience*, 12(1):85–102, 1989.

[82] Rosel M Mulkey and Robert C Malenka. Mechanisms underlying induction of homosynaptic long-term depression in area ca1 of the hippocampus. *Neuron*, 9 (5):967–975, 1992.

[83] Alain Artola and Wolf Singer. Long-term depression of excitatory synaptic transmission and its relationship to long-term potentiation. *Trends in neurosciences*, 16(11):480–487, 1993.

[84] Bernabé Linares-Barranco and Teresa Serrano-Gotarredona. Memristance can explain spike-time-dependent-plasticity in neural synapses. *Nature proceedings*, 1:2009, 2009.

[85] Yang Dan and Mu-Ming Poo. Spike timing-dependent plasticity: from synapse to perception. *Physiological reviews*, 86(3):1033–1048, 2006.

[86] Natalia Caporale and Yang Dan. Spike timing-dependent plasticity: a hebbian learning rule. *Annu. Rev. Neurosci.*, 31:25–46, 2008.

[87] Adria Bofill-i Petit and Alan F Murray. Synchrony detection and amplification by silicon neurons with stdp synapses. *IEEE Transactions on Neural Networks*, 15 (5):1296–1304, 2004.

[88] Hualiang Zhuang, Kay-Soon Low, and Wei-Yun Yau. A pulsed neural network with on-chip learning and its practical applications. *IEEE Transactions on Industrial Electronics*, 54(1):34–42, 2007.

[89] Thomas Jacob Koickal, Alister Hamilton, Su Lim Tan, James A Covington, Julian W Gardner, and Tim C Pearce. Analog vlsi circuit implementation of an adaptive neuromorphic olfaction chip. *IEEE Transactions on Circuits and Systems I: Regular Papers*, 54(1):60–73, 2007.

[90] Jordi Cosp and Jordi Madrenas. Scene segmentation using neuromorphic oscillatory networks. *IEEE Transactions on Neural Networks*, 14(5):1278–1296, 2003.

[91] Saber Moradi and Giacomo Indiveri. An event-based neural network architecture with an asynchronous programmable synaptic memory. *IEEE transactions on biomedical circuits and systems*, 8(1):98–107, 2014.

[92] Paul A Merolla, John V Arthur, Rodrigo Alvarez-Icaza, Andrew S Cassidy, Jun Sawada, Filipp Akopyan, Bryan L Jackson, Nabil Imam, Chen Guo, Yutaka Nakamura, et al. A million spiking-neuron integrated circuit with a scalable communication network and interface. *Science*, 345(6197):668–673, 2014.

[93] Janardan Misra and Indranil Saha. Artificial neural networks in hardware: A survey of two decades of progress. *Neurocomputing*, 74(1-3):239–255, 2010.

[94] Ben Varkey Benjamin, Peiran Gao, Emmett McQuinn, Swadesh Choudhary, Anand R Chandrasekaran, Jean-Marie Bussat, Rodrigo Alvarez-Icaza, John V Arthur, Paul A Merolla, and Kwabena Boahen. Neurogrid: A mixed-analog-digital multichip system for large-scale neural simulations. *Proceedings of the IEEE*, 102(5):699–716, 2014.

[95] Ahmad Muqeem Sheri, Aasim Rafique, Witold Pedrycz, and Moongu Jeon. Contrastive divergence for memristor-based restricted boltzmann machine. *Engineering Applications of Artificial Intelligence*, 37:336–342, 2015.

[96] Wolf Lukas Hellweg. Memristor-based artificial neural network for unsupervised pattern learning and recognition. Master's thesis, Institute of Nano- and Medical Electronics, Technical University of Hamburg, 2016.

[97] Ranjan Ranjan, Alexandros Kyrmanidis, Wolf Lukas Hellweg, Pablo Mendoza Ponce, Lait Abu Saleh, Dietmar Schroeder, and Wolfgang H. Krautschneider. Integrated circuit with memristor emulator array and neuron circuits for neuromorphic pattern recognition. In *39th International Conference on Telecommunications and Signal Processing (TSP 2016)*, Vienna, Austria, 27-29 June 2016.

[98] Bernhard Nessler, Michael Pfeiffer, and Wolfgang Maass. Stdp enables spiking neurons to detect hidden causes of their inputs. In *Advances in neural information processing systems*, pages 1357–1365, 2009.

[99] Thomas Martinetz and Klaus Schulten. Topology representing networks. *Neural Networks*, 7(3):507–522, 1994.

[100] Philine Wangemann and Jochen Schacht. Homeostatic mechanisms in the cochlea. In *The cochlea*, pages 130–185. Springer, 1996.

[101] Gina Turrigiano. Too many cooks? intrinsic and synaptic homeostatic mechanisms in cortical circuit refinement. *Annual review of neuroscience*, 34:89–103, 2011.

[102] Abhronil Sengupta, Priyadarshini Panda, Parami Wijesinghe, Yusung Kim, and Kaushik Roy. Magnetic tunnel junction mimics stochastic cortical spiking neurons. *Scientific reports*, 6:30039, 2016.

[103] Damien Querlioz, Olivier Bichler, Philippe Dollfus, and Christian Gamrat. Immunity to device variations in a spiking neural network with memristive nanodevices. *IEEE Transactions on Nanotechnology*, 12(3):288–295, 2013.

[104] Maruan Al-Shedivat, Rawan Naous, Gert Cauwenberghs, and Khaled Nabil Salama. Memristors empower spiking neurons with stochasticity. *IEEE Journal on Emerging and Selected Topics in Circuits and Systems*, 5(2):242–253, 2015.

[105] Sheng-Feng Yen, Jie Xu, Manu Rastogi, John G Harris, Jose C Principe, and Justin C Sanchez. A biphasic integrate-and-fire system. In *Circuits and Systems, 2009. ISCAS 2009. IEEE International Symposium on*, pages 657–660. IEEE, 2009.

[106] Xinyu Wu, Vishal Saxena, Kehan Zhu, and Sakkarapani Balagopal. A cmos spiking neuron for brain-inspired neural networks with resistive synapses and in situ learning. *IEEE Transactions on Circuits and Systems II: Express Briefs*, 62(11): 1088–1092, 2015.

[107] Phillip E Allen, Douglas R Holberg, Phillip E Allen, and PE Allen. *CMOS analog circuit design*. Holt, Rinehart and Winston New York, 1987.

[108] Prof. Dr.-Ing. Wolfgang Krautschneider. Circuit design. *Circuit design course*, chapter 2:page 3.

[109] Behzad Razavi. *Design of analog CMOS integrated circuits*. Tata McGraw-Hill, 2010.

[110] Yannis Tsividis and Colin McAndrew. *Operation and Modeling of the MOS Transistor*. Oxford Univ. Press, 2011.

[111] Christopher D Moyes and Patricia M Schulte. *Animal Physiology*. Benjamin Cummings San Francisco, CA, 2005.

[112] Wikipedia contributors. Refractory period physiology wikipedia the free encyclopedia. 2018. Online; accessed 6-September-2018.

[113] Jan Benda and Andreas VM Herz. A universal model for spike-frequency adaptation. *Neural computation*, 15(11):2523–2564, 2003.

[114] Ying-Hui Liu and Xiao-Jing Wang. Spike-frequency adaptation of a generalized leaky integrate-and-fire model neuron. *Journal of computational neuroscience*, 10(1):25–45, 2001.

[115] Bard Ermentrout, Matthew Pascal, and Boris Gutkin. The effects of spike frequency adaptation and negative feedback on the synchronization of neural oscillators. *Neural Computation*, 13(6):1285–1310, 2001.

[116] Erkki Oja. Simplified neuron model as a principal component analyzer. *Journal of mathematical biology*, 15(3):267–273, 1982.

[117] Mirko Hansen, Finn Zahari, Hermann Kohlstedt, and Martin Ziegler. Unsupervised hebbian learning experimentally realized with analogue memristive crossbar arrays. *Scientific reports*, 8(1):8914, 2018.

List of Figures

List of Tables

Curriculum Vitae

Personal information

Name: Rajeev Ranjan

Date of birth: 21.09.1984

Nationality: Indian

Studies

08.2014-05.2018:   Scientific Employee at Hamburg University of Technology
                   Institute of Nano- and Medicine Electronics, TUHH, Hamburg
                   Project : Memristor

10.2012-08.2014:   Master of Science in Microelectronics and Microsystems
                   Hamburg University of Technology (TUHH), Hamburg

06.2004-06.2008:   Bachelor of Technology in Electrical and Electronics Engineering
                   SASTRA University, Thanjavur, India.

Work Experience

06.2009-06.2011:   Analog Layout Design Engineer
                   Cosmic Circuits, Bangalore, India.

06.2011-06.2012:   Senior Engineer
                   Mindtree Pvt. Ltd deployed at Intel, Bangalore, India.

Bisher erschienene Bände der Reihe

Wissenschaftliche Beiträge zur Medizinelektronik

ISSN 2190-3905

10   Jan Claudio Loitz   Novel Methods in Electrical Stimulation with Surface
                              Electrodes
                              ISBN 978-3-8325-4885-8   45.00 EUR

Alle erschienenen Bücher können unter der angegebenen ISBN-Nummer direkt online
(http://www.logos-verlag.de) oder per Fax (030 - 42 85 10 92) beim Logos Verlag
Berlin bestellt werden.